A Bit of Controversy

Pat Hawker - A Radio Life

by

Steve White, G3ZVW

Published by the Radio Society of Great Britain, 3 Abbey Court, Fraser Road, Priory Business Park, Bedford MK44 3WH.

Web site: www.rsgb.org

Tel: 0870 904 7373

Fax: 0870 904 7374

ISBN: 9781-9050-8640-5

Cover design: Dorotea Vizer, M3VZR

Proof reading: George Brown, MW5ACN

Production: Mark Allgar, M1MPA

The opinions expressed in this book are those of the author and not necessarily those of the RSGB. While the information presented is believed to be correct, the author, publisher and their agents cannot accept responsibility for consequences arising from any inaccuracies or omissions.

Printed by Latimer Trend & Company Ltd of Plymouth, England

Contents

Foreword

In undertaking the role of biographer, one never quite knows what will be discovered about the person to be written about, even if they are already an extremely familiar character.

This has certainly turned out to be the case for me, because although I knew Pat Hawker had written 'Technical Topics' for RSGB's *RadCom* for almost fifty years, that he had also been part of the Secret Wireless War, had seen service abroad and was responsible for other publications, until I researched this book I had no idea just what a prolific author he had been or that he had literally helped shape the world we live in today.

In the pages that follow you will learn of Pat's background as the son of a master baker, of his childhood homes, schooling, friends, qualifications and early career. Radio came early to Pat, indeed he was at just the right age to be bitten by the 'bug' when short wave broadcasting was in its heyday. Within a few months of WWII being declared, he started to play a part. It was one that would develop and lead to a change of career after the hostilities, indeed Pat's interest in radio led to him to spending practically his entire working life in communications and broadcasting. He became the highly respected editor of numerous books and a contributor to many magazines, but it didn't always keep him out of trouble.

In practically every field of human endeavour a few people stand out. Pat Hawker is undoubtedly one of those people. His life and times have been spent almost entirely in radio in one way or another, and as a result he has an encyclopaedic knowledge of the subject and the people. His service to his country, the radio and television industry and to amateur radio have been exemplary, indeed they may never be equalled.

You may feel that you already know Pat, indeed countless thousands already do, but by the time you reach the end of this book you will almost certainly realise that you didn't know all there was to know.

Steve White, G3ZVW

The Early Years

Pat's father, William, was born in 1884 in the town of Dunster, near Minehead in Somerset. His profession was as a Master Baker, indeed he ran a bakery, confectionery and tea room in Friday Street, Minehead, from 1904 to 1938. It wasn't a terribly successful business and during the depression of the 1930s the family were not well off. His mother, Norah the seventh child of a seventh child, was born in Windsor in 1888. Her father and brothers ran antique businesses in various towns and countries, while her sisters ran a high-fashion dress shop in Knightsbridge, London.

Although Pat's parents were very different and not really well suited, they were both good parents in their own way. Pat was much closer to his mother and later regretted that he did not always give his father – who died just after the end of WWII – the understanding and respect he deserved. Pat's father would have been better suited to farming than baking, but things were better after he gave up the baking business and retired in 1938. In this year the family moved to Redfields, a pleasant house in Whitecross Lane, which is still

Pat's mother, Norah.

in Minehead. It lies on the slopes of North Hill, which overlooks the town.

Mr and Mrs Hawker's first child was Betty, who was born in 1910, twelve years before Pat. From about 1925 to 1932 she assisted her parents in the family business, but she also appeared in films. In 1929 she played the part of a bridesmaid in the Alfred Hitchcock film *The Farmer's Wife*, which was shot in Elstree studios and on location in West Somerset. Later she acted as a double to Czech-born star Anna Ondra in car chase scenes. She moved to London in about 1933 and worked in Fortnum and Masons, Piccadilly, until she married an architect in 1934.

Next came Ryland (Roy), who was five years older than Pat. He was the athletic and outgoing member of the family. Whilst at school he became the School Captain, and was also Captain of the Rugby XV. He was awarded a scholarship to the Royal Military Academy and subsequently served first in the British Army, then the Indian Army. He had a talent for foreign languages and learned Urdu, Pashto and Arabic. He saw service on the North-West Frontier, in Greece, the Greek islands, Turkey and Afghanistan (as the British Military Attaché). Roy was awarded an MBE in the 1953 Queen's Birthday Honours List and a "Certificate of Achievement for Exceptionally Outstanding Performance of Duty" for his work as Deputy Assistant Adjutant and Quartermaster General at Eriforce HQ during the thinning-out and final evacuation of the British garrison in Eritrea. Interested in all forms of wildlife, he is described by his daughter as a "very complex and private man, who in later life did not suffer fools gladly".

Pat's immediate elder brother is Peter, who is 3½ years older than him. Peter's schooldays must have been rather overshadowed by Roy's successes, but he was a keen Cub and enjoyed a cheerfully mischeivous time at school. Nicknamed Pug, his skills were more practical than academic. Along with his sister Betty he also appeared in a 16mm semi-professional local film, both playing the roles of gypsies, while Roy provided a pet red squirrel for it. Peter trained in agriculture and farming at the Somerset Farm Institute. In early 1939 he volunteered to train as an RAF Pilot Officer, but failed his navigation course so left. At the outbreak of WWII he volunteered for service in the Royal Army Veterinary Corps, seeing service in Palestine, the Western Desert, and then an ill-fated expedition to assist the Greeks. In 1941 he was taken prisoner of war by the Germans. He remained a PoW for some four years in Germany and Austria, making some unsuccessful attempts to escape but also spending time working on an Austrian farm. At the end of WWII, Peter was one of the last PoWs to be repatriated, causing much anxiety to his mother as several weeks passed by without news of him. After the war he returned to Somerset, married, had a family, and continues to live there, now in a care home.

Last of the Hawker children is Pat, who was born on 5 April 1922, the first day of the new financial year. His earliest memories are of 1924 or 1925, at which time the family employed Enid, a young live-in maid/nurse-maid. Enid took Pat for a short walk, meeting her 'follower', who was a waiter at the Plume of Feathers Hotel. They took Pat on a mini cruise on one of Campbell Line's 'White Funnel' paddle steamers that used to plow between the coastal towns of the Bristol Channel. On this day they returned home several hours later than expected, by which time Pat's mother was in a state of acute anxiety as to what had happened to him. The result of this is that Pat's much-loved Enid disappeared from his life. Subsequently he would often play, unseen by customers, behind the counter of the shop, learning the alphabet from the writing on the large tins of biscuits.

A family day out with his brothers, sister and father on one of the White Star liners that used to steam back and forth across the Bristol Channel, between Mine-head and Barry. Pat is second from the left.

Apart from himself, Peter is the only one still alive.

Schooling

At the age of five, Pat attended a few terms at Madame Pine-Gilbert's small 'branch' mixed kindergarten, which was an independent private school. The main school was in Taunton, the county town of Somerset. The Minehead branch consisted of about ten children, taught by a young teacher. It must all have proved uneconomic, as the establishment soon closed.

Pat then attended the Modern School (boys), run by a Mr Richards, who came from Wales. Pat's brothers had been there and it had grown to quite a reasonable size, but due to the building of the new Minehead County School in 1928 it too closed as

A Bit of Controversy

Mr Richards (as a geography teacher) and most of the pupils soon transferred there. Mr Richards' wife continued to run a small junior school of about eight pupils, which Pat remembers as being quite good. He recalls going outside – possibly in 1929 – to see the Graf Zeppelin pass over on one of its transatlantic voyages. Some years later he also saw the ill-fated Hindenburg airship returning from the USA.

Pat started attending the Minehead County School in September 1932, by which time there were about 200 pupils there. A co-educational school, Pat describes it as "very good". He started in the 'B' stream, but was transferred to the 'A' stream after just one term. The County took fee paying students as well as students on scholarships from council schools in local villages. They were generally quite bright people and Pat was about 18 months below the average age of his class, although he's not sure if that was an advantage or not.

A scene from a film which Minehead residents are making with the aid of local Boy Scouts. The players are (left to right): Pat Hawker, Scout Jim Millord, C. T. Chapman, and W. R. Webber. (Photo: Martin Cross.)

In about 1933, whilst at school, Pat played a role in the last of the Martin Cross 16mm silent films, *Trail of Youth*. He remembers two things clearly about the film. Firstly, it could still be rented from the Kodak Film Library after the War. Secondly, it convinced him that he was no actor (although he did once play the role of The Duke of Venice in Shakespeare's *Merchant of Venice* in the school play).

The school exam structure in the 1930s was different from

today. Most pupils left after taking the Oxford Senior Certificate examinations at the end of Form V, with a few staying on for a year or two in Form VI. From about 1936 a handful of pupils would take the Bristol Higher School Certificate examinations at Bristol University. More took the Civil Service exams for Clerical or Executive grades.

Pat ended-up taking his School Certificate examinations at the age of 15. These were the equivalent of today's O-levels. This was followed by the Bristol Higher School Certificate examinations – the equivalent to today's A-levels – at the age of 17, whereas most people took them at 18. These exams had to be taken in Bristol itself, not at the local school. Physics, Pure Mathematics and Applied Mathematics were Pat's subjects.

From a scene in the film, Pat, right, on board the 'ship' where he was held after being 'kidnapped'.

One area in which Pat did not excel was sport and athletics, although he enjoyed swimming. He has not taken part in any active sports since leaving school. Despite not being physically active, he has enjoyed reasonably good health. The worst things that happened have been acne in his youth and concussion from falling off his bicycle.

Always the tall one, Pat is standing in the centre of the back row in this photo of school prefects.

First interest in radio

It was during his years of schooling that the radio 'bug' bit Pat. When an uncle went to live with them for a short while in about 1927, he took a radio with him. This would have been the first time Pat really experienced wireless broadcasting.

In fact Pat wasn't the first member of the Hawker family to express an interest in radio, it was his older brother Roy. In about 1930 Roy had built a crystal set to receive the low power BBC Cardiff transmitter as it then was. Apparently it didn't work particularly well. After this, through the columns of *Exchange and Mart*, Roy bought sufficient Black Cat cigarette coupons to claim a KB Kitten battery receiver.

Shortly afterwards, the BBC started building twin stations at Washford Cross, which is about seven miles from Minehead. When

it opened in 1934 it put a very strong signal into Minehead. It carried National and Regional Services, but also a lot of programmes in Welsh, because Washford is just across the Bristol Channel from Wales. It was positioned where it was specifically because it would put a good signal into all of South Wales. "Programmes in Welsh didn't go down too well in West Somerset", remembers Pat, who by this time was already building crystal sets of his own.

Walking along the esplanade at Minehead, Pat and his friend Gordon Wood, who became 2DNZ.

As the construction phase developed, so did the magazine buying. Pat bought *Amateur Wireless* (3d weekly) until it merged with *Practical Wireless* and became known as *Practical & Amateur Wireless*. He also bought *Wireless Magazine* (1s, monthly) and later still *Television & Short Wave Magazine* (also monthly). Occasionally he also bought *Wireless World* or the BBC magazine *World Radio* (both weekly). Partly for Pat's benefit, his mother would buy the non-technical *Radio Pictorial* (weekly).

Building crystal sets evolved into building one with a simple 1-valve audio amplifier. The valve, which has its filament powered by a 2-volt battery, cost about 5/6d, which was a considerable sum of money in its day. Pat financed its purchase mostly from money he earned selling fresh bread rolls early in the morning to the numerous Bed and Breakfast establishments in Minehead, a quiet seaside resort and 'Gateway to Exmoor' in the 1930s, before the postwar arrival of Butlins Holiday Camp.

Then, in 1935, came a 2-valve short wave TRF receiver. The 2-valve radio was a very simple design. No soldering was involved and it incorporated the inevitable cardboard toilet roll as a coil former. The HF broadcast stations Pat heard included Zeesen, Rome, Schenactady, Pittsburgh and Melbourne. Using his bed-frame as the antenna, one night he heard somebody talking in English. Pat didn't know it at the time, but he was listening to his first radio amateur – LA1G. With no test equipment, it would have been difficult to determine what frequency band was being listened to, let alone what frequency, but it turned out to be the 20m band. Pat soon found the 40m band and heard British stations talking to one another.

In 1936, along with school friends, a small radio club was formed at Minehead County School. There were about eight or nine members. Charles Bryant and Pat got their fathers to take out Artificial Aerial – or 'AA' – licences, which were seen as a stepping stone to the full licence. In those days, nobody under the age of 21 could be a licence holder, so the parents of keen youngsters invariably held licences and acted on behalf of their offspring, designated as 'their agents'. No exams needed to be taken, but there were lots of forms to fill in. Pat became 2BUH and Charles 2BXZ. Other AA licences followed. Gordon Wood became 2BNZ, John Mansell became

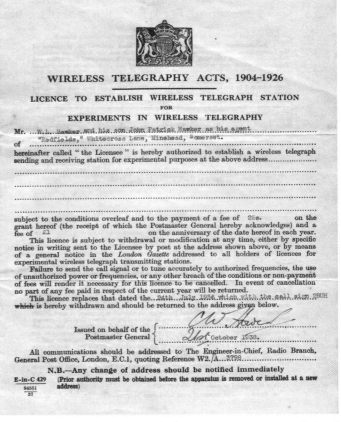

WIRELESS TELEGRAPHY ACTS, 1904–1926

LICENCE TO ESTABLISH WIRELESS TELEGRAPH STATION
FOR
EXPERIMENTS IN WIRELESS TELEGRAPHY

Mr. ...W.L...Hawker...and...his...son...John...Patrick...Hawker...as...his...agent...............
of ..."Redfields," Whitecross Lane, Minehead, Somerset.
hereinafter called "the Licensee" is hereby authorized to establish a wireless telegraph sending and receiving station for experimental purposes at the above address...............

subject to the conditions overleaf and to the payment of a fee of 25s. on the grant hereof (the receipt of which the Postmaster General hereby acknowledges) and a fee of £1 on the anniversary of the date hereof in each year.
    This licence is subject to withdrawal or modification at any time, either by specific notice in writing sent to the Licensee by post at the address shown above, or by means of a general notice in the *London Gazette* addressed to all holders of licences for experimental wireless telegraph transmitting stations.
    Failure to send the call signal or to tune accurately to authorized frequencies, the use of unauthorized power or frequencies, or any other breach of the conditions or non-payment of fees will render it necessary for this licence to be cancelled. In event of cancellation no part of any fee paid in respect of the current year will be returned.
    This licence replaces that dated the..24th..July..1936..which..with..the..call..sign..2BUH which is hereby withdrawn and should be returned to the address given below.

Issued on behalf of the { ....
Postmaster General { 21st October 1938.

    All communications should be addressed to The Engineer-in-Chief, Radio Branch, General Post Office, London, E.C.1, quoting Reference W2./A../.3798.
    N.B.—Any change of address should be notified immediately
E-in-C 429   (Prior authority must be obtained before the apparatus is removed or installed at a new
84551   address)
37

The original licence, issued to Pat via his father in 1938.

BRITISH EXPERIMENTAL HIGH FREQUENCY STATION.

2BUH

QRA

"Redfields,"
White Cross Lane,
MINEHEAD.

To RADIO................ ...... Ur Sigs Fone/CW QSA........hrd hr................ ....

at...............GMT. R........T........QRG...........MC. Modulated........%

QSB to R........QRM...Rx.............-V-............

Remarks : (see over)

AERIAL
73's es FB DX OM
de J. P. HAWKER.

PSE QSL via R.S.G.B. TNX OM.   Opr 2BUH

Printed by 2GDT ELLAND.

Pat's QSL card from his days as an 'Artificial Aerial' licence station.

A Bit of Controversy

2CMN and John Payne became 2BYK. Payne subsequently became G2BYK for the rest of his life.

Pat remembers the club visiting the BBC transmitter site at Washford Cross and being shown around. At around this time, Pat and Charles got interested in home recording. The only kind that could be done in those ways was on aluminium discs, which had to be played with a fibre needle. Pat had first come across this system on a visit to the Radiolympia exhibition in London, where visitors could put sixpence in a slot and record their own voice. On a school Open Day a recording service was offered, but not one parent took advantage of the facility so the students had to record themselves. Pat remembers the tracking as being very difficult.

Pat readily acknowledges that Charles Bryant was far more socially minded than himself. Charles got to know an amateur at Porlock, about six miles west of Minehead. Pat also went over and met him. Occasionally Pat and Charles would go to Taunton, the county town of Somerset. It lies some 24 miles south east of Minehead, the attraction being that it had quite a large radio society. Pat remembers G5AK, G2LM who was a doctor and the former Mayor of Taunton (he used a Morse operator, rather than pound the key himself) and G6LQ. They attended a few meetings and both joined RSGB in the autumn of 1936. Joining the Society was a somewhat more complicated business in those days, because you had to be vouched for by existing members. Being the more outgoing one, Charles got someone to vouch for him. As soon as he was admitted, he vouched for Pat. Consequently there was a two-month gap between when they became members.

In 1936 Pat had his first article about short wave listening published, in his school magazine.

Pat's first 'transmitter', used in his 'AA' days as 2BUH, was nothing more than an oscillator, although he had experimented with oscillators before that. At one time Pat and friends ran a little wireless system within his Friday Street home. It enabled him to talk between his bedroom and a loft, which was about 100ft away.

Other Interests

Although radio became Pat's main hobby, he was also interested in collecting stamps, cigarette cards and going to the cinema. Pat was keen on the weekly magazine *Skipper*, one of the DC Thompson stable of publications (others being *Wizard*, *Rover* and *Hotspur*). An early member of the newly-established Minehead Public Library, it wasn't long before Pat began reading the many

Edgar Wallace novels, P G Wodehouse and the Ashenden (intelligence) stories of Somerset Maugham. He also remembers reading *Greek Memories*, Compton McKenzie's account of his work in the Secret Service in WWI and for which he was prosecuted for breaching the Official Secrets Act.

As well as elementary radio books, Pat became interested in codes and ciphers. His favourite radio storyteller was A J Alan, who later turned out to be Leslie Lambert, G2ST, of the Government Code and Cipher School. At one time Lambert had been a professional children's magician, but he died in 1941 while working at Bletchley Park - by which time Pat had become involved with such intelligence matters through the Radio Security Service.

Real transmitting licences!

Charles and Pat began to learn Morse, with Charles passing his GPO Morse Test in 1938 at the age of 16. He became G3SB and is still active today as GW3SB. Pat took his Morse Test at his local Post Office on 30 September 1938, at the height of the Munich Crisis, and received the callsign G3VA in late October 1938. They had both benefited from Morse training sessions (for a modest fee) from an aged, retired Royal Navy wireless telegraphist that Charles had found.

Still 2BUH, Pat stands outside his house with a display of amateur radio equipment.

It was around this time that Pat began to have features published in the amateur radio press:

A Simple 1.7 Mc/s Converter	*Short Wave Magazine*, 1938
Simple Frequency Meter	*T&R Bulletin*, 1939

For actual transmitting a quartz crystal was essential in the 1930s, indeed you had to send the Post Office a 'crystal certificate' to demonstrate that you had a calibrated crystal and that it was within an amateur band before they would issue a licence. In those days the only amateur bands were 1.7MHz, 7MHz and 14MHz, so the standard practice was to frequency double from 7MHz if you wanted to transmit on 14MHz. For the 1.8MHz band it was necessary to provide the Post Office with a second certificate, as you couldn't use the same crystal to operate on it. G2NH owned the Quartz Crystal Company, from which you could buy suitably calibrated crystals, but Pat borrowed Charles' 1.7MHz crystal certifi-

A Bit of Controversy

QRA- "REDFIELDS," WHITECROSS LANE,
MINEHEAD, SOMERSET, ENGLAND.

To Radio..Ur.................................sigs wkd here
...................193...at.................GMT/BST.................MC
RST..QRM.................QRN................
Transmitter..Ant.................................
Receiver..Input...........Watts

PSE QSL OM. 73 es DX de J. P. Hawker Opr (ex 2BUH)
PRINT G5KT BRISTOL 6.

cate to get his permit to transmit on the band. His first ever contacts were on telephony, although he had already experimented with another friend, Gordon Wood, who became 2DNZ, on what they thought was about 5 metres.

By this time Pat's parents had moved from the shop in the town to the house on the hill overlooking it. For their VHF experiments the only way Pat and Charles could estimate the frequency was to have an aerial about 8ft long with a little light bulb in the middle of it. When feeding RF into it, if the bulb lit you knew you were somewhere in the vicinity of 60MHz. Leaving the transmitter running with a clock next to the microphone, they had a portable super-regenerative receiver that they could carry around the neighbourhood and listen to discover how far away they could hear the ticking. Pat remembers being able to hear the clock up to a mile away, up one of the valleys, but not very strongly.

A mains transformer was purchased from the Quartz Crystal Company, which provided Pat with a 350V supply. He used it to destroy several receiving valves. Incidentally, the construction of equipment in those pre-WWII days was an open rack and panel type. By today's standards you would hardly call what Pat had a 'station', but in 1939 with a two-valve receiver and a two-stage (crystal oscillator and T20-valve power amplifier) he contacted amateurs in North and South America, Southern Rhodesia and Australia on 14MHz CW, many British and European stations on 7MHz CW and phone, and local West Country stations on 1.7MHz.

The War Years

Leaving School

Pat had taken his final school exams in about June 1939, with the summer school term ending in July. He left without knowing if he had passed the exams and with no clear idea of what he was going to do in life. During the spring he went to London and attended an interview at the GEC research laboratories, to see if they would offer him a research apprenticeship, but they didn't.

31 August
     Pat had four QSOs. They were with LY1AP in Lithuania, YR5BV in Romania, LA8J in Norway and SM6QN in Sweden. That evening Pat heard something on the broadcast radio that led him to make the following note in his log book. "During the BBC news bulletin at 2100 heard announcement that licences have been withdrawn. All valves removed from transmitter. Further information awaited." Although war had been very much anticipated, Pat was still enjoying the summer holidays at the time and little guessed that three of the four countries he had just enjoyed contact with would be ravaged by that war.

2 September
     Pat wrote in his log book, "GPO engineers removed transmitter". Soon, Defence Regulations made it illegal to be in possession of valves of more than 10 watts dissipation, quartz crystals or any form of transmitting equipment. It wouldn't be until February 1946 that Pat had his next 'official' amateur radio QSO, although there were certainly some who jumped the gun, concocted fictitious

callsigns and resumed activity immediately the war in Europe ended in May 1945 - some even before!

3 September

Britain declares war on Germany, but the amateur bands don't fall immediately silent. Although subjected to increasing restrictions, American amateurs in particular remained active until the Japanese attack on Pearl Harbor in December 1941. More importantly, although transmitting licences were withdrawn here and transmitters impounded, receivers were not. The British were still listening, and although it was not publicised they were listening in great numbers. Pat bought the final disc of a set of 78rpm Morse training records issued by Columbia, gradually increasing his speed to about 25wpm.

At around this time Pat's father had been speaking with his solicitor, who suggested that accountancy was a good profession, so in mid September, shortly after war was declared, Pat took out an articled clerkship with a local chartered accountants. He was to work there until November 1941 and again from September 1946 to August 1947.

Becoming a Volunteer Interceptor

In about April 1940 Pat received at home a curious letter from a Lord Sandhurst, asking if he would be prepared to do some voluntary work for the government. If he wished to express an interest and know more, he was to sign the enclosed extract from the Official Secrets Act and return it to PO Box 385, Howick Place,

Lord 'Sandy' Sandhurst (centre).

London SW1. This Pat duly did. It wasn't long before Pat discovered that his friend Charles Bryant, G3SB, had also received a copy of the same letter. Having left school a year before Pat, Charles was training as a solicitor at the time – and being a trainee solicitor he wasn't about to sign anything until he knew what he would be getting himself into! Some correspondence ensued and after a while Charles too became a Voluntary Interceptor ('VI'), remaining one throughout the war. Pat and Charles were just two of more than a thousand radio amateurs who were approached during the war. Lord Sandhurst had consulted Ken Alford, G2DX, then Arthur Watts, G6UN, the then President of RSGB, who recruited many London amateurs.

Within a couple of weeks of replying to his invitation, Pat received a booklet that showed what sort of preambles the British and German services used. They didn't want Pat to bother listening to service traffic, but to concentrate on anything unusual. The brief was to write down anything he heard, log it and send it in. Unfortunately the only receiver Pat had access to at the time was a Philco radiogram in the family's living room, to which Pat had added a BFO to resolve CW. The calibration was such that you didn't know within 50kHz or so what frequency you were listening to. To listen more effectively Pat was lent a Hallicrafters S20R, a receiver known as the Sky Champion. Pat used the S20R for a year or so and describes it as being quite a good receiver in its day. Incidentally, the reason Pat was lent an American receiver is that there was hardly any suitable British equipment available as Britsh factories were desperately replacing signals equipment lost at Dunkirk, whereas the American amateur radio scene was vibrant and several companies were producing communication receivers.

The Hallicrafters S20R, also known as the 'Sky Champion'.

The organisation Pat was now working for was the Radio Security Service. For an 18-year-old it must have been quite thrilling to know that he was part of what appeared to be a secret service, although originally those in RSS knew little of the background to the organisation or what it was. Even though the RSS had established a South West Regional Office at 27 Dix's Field, Exeter, all they knew was that it was associated with the secret service. In fact it was part of the War Office, who also ran the Y service – the official intercept service – but as a separate section to it. It was in fact MI8c, although this was never mentioned at the time.

RSS remained a separate and autonomous organisation throughout the war, although it later merged and came under the control of Colonel (later Brigadier) Richard Gambier-Parry (ex G2DV) in charge since 1938 of the Communications Section (Section VIII) of the Secret Intelligence Service (SIS/MI6) that in wartime became the first Special Signals Unit (SSU1) and then in 1941 the Special Communications Unit (SCU).

As a VI, Pat knew nothing of the fact that RSS had three intercept stations, each of which had been set up by the Post Office. One was near Forfar in Scotland, one in Northern Ireland and one in Cornwall. The GPO was responsible for recruiting and training the operators. Some were former Merchant Navy Radio Officers, but less successful were their attempts to train Post Office counter staff, who weren't particularly successful as 'general search' people. As a result, at one time the War Office actually wanted to close down the entire RSS operation and absorb it into the Y service.

The photo taken of Pat for his RSS identity card.

The problem was that the original purpose of the operation was to listen for German agents in Britain, be they communicating with Germany or operating beacon stations. In fact the Germans could navigate sufficiently well, so they didn't need clandestine beacons anyway. Originally the Post Office set up a large network of mobile direction finding units to chase down any enemies within, but Pat feels the exercise was all pretty useless.

Upon reaching the age of 19, in April 1941, Pat applied to join the RAF. Unfortunately he has always suffered from what a doctor once called "a nervous stomach", which manifests itself by him having a poor night's sleep and suffering a raised temperature before any major event he attends. When he was summoned to Exeter for a medical, needless to say it was found that his temperature was high. Pat was told to go and sit outside for a few hours, then come back. Later that day the result was the same, and the same again a month or so later, and yet again when he was called-up and took a medical at Taunton. Consequently he continued his day job in the accountant's office and being a VI during the evenings.

The 'nervous stomach' problem would remain with Pat throughout his life. First broadcasts, television appearances, lectures at professional conferences or radio clubs were all an ordeal, although it was gradually overcome for repeated events.

By that time RSS had moved its operation to the northern edge of London. The anonymous postal address was PO Box 25, Barnet. By June 1940 it had discovered that the stations the VIs were picking up and logging were in fact Abwehr stations – the

German Secret Service. By that time Major Gill and Trevor Roper – who became Lord Dacre – at Box 25 were able to decode some of this traffic and proved that some of what the VIs had received was important to the Intelligence and Security Services. This caused a big fuss, because originally Bletchely Park had refused to accept VI intercepts, mistakenly believing them to be stations in Russia. It wasn't until it had been proved beyond doubt that these were intelligence messages – including spy ships off the Norwegian coast – that Bletchley Park was finally persuaded to set up its own unit under Oliver Strachey to handle the RSS messages. They were soon able to decode the quite difficult hand codes used by the Abwehr. It took them another year to decode the Abwehr Enigma traffic, but this was also finally achieved.

The Abwehr

Before the war, the German government had tried to 'nazify' the German amateur radio movement. Later they came to regret this. Goering, as Chief of the Luftwaffe, is on record as saying in 1943; "We smashed-up the amateur radio 'ham' clubs and wiped them out, and we made no effort to help those thousands of small inventors. Now we need them." It was on the orders of Adolf Hitler himself that this destructive activity had been undertaken, because amateur radio might have been able to provide communication facilities for disaffected groups.

Wilhelm Canaris, leader of the Abwehr until 1944. He was later executed for his part in the July 1944 attempt to assassinate Adolf Hitler.

Something the Abwehr didn't know is that in the summer of 1939 a German suitcase radio that had been delivered to Victoria Station in London had been obtained by the Security Service (MI5). It should have been delivered to Arthur Watts, a Welsh electrical engineer who, in the 1930s, had worked for both British Intelligence and the Abwehr. With Watts' radio, signal plan and codes, a VI operator made contact with the Abwehr control station at Hamburg, run by Major Trautmann. Spurred-on by this success, the VI noted that Hamburg was contacting other stations, including what turned out to be a clandestine ship sailing in Norwegian waters. Other VIs began to report similar traffic, which enabled the RSS to start working out how the Abwehr network operated and, importantly, the start of the double-cross operation, 'playing back' the German agents as they arrived.

Although part of the German army, the Abwehr operated its

communications entirely separately, just as the British Secret Service did. As the Germans set up communication bases in Norway, Belgium, Paris, and the west coast of France, RSS kept abreast of the situation. The Abwehr also established a busy communication station working in Sofia, the Balkans, plus after they took over Greece in 1941, Athens and the Greek islands. Their main bases were in Berlin, Hamburg, Weisbaden and Vienna. The Abwehr radio networks became a very large organisation, including many German radio amateurs. Oddly, after an initial suspension of all amateur radio activity in Germany at the outbreak of war, some were allowed back on the air. This was partly for propaganda purposes, to demonstrate that everything was 'normal', but there were also links with the Abwehr.

The French Special Services under the organisation of Gustave Bertrand were undertaking parallel work to RSS, intercepting traffic from the Weisbaden control station and "playing back" the radio of a controlled German agent. After the fall of Norway and France, the German networks expanded rapidly, until they covered practically all of Europe and North Africa, while the Hamburg station also worked to agents in North and South America – some genuine, some controlled by the FBI.

Pat's brother Peter (left) and another Prisoner of War on a farm in Austria. Peter was captured in 1941 and remained a prisoner until after hostilities.

For some reason the Abwehr had decided that none of their non-traffic conversations should be conducted in German. Instead, and perhaps not surprisingly, it was in amateur radio format and liberally sprinkled with amateur type abbreviations. Amongst other things, the German operators would often give their initials. Needless to say RSS in Barnet soon built up a large database of the initials, and this turned out to be very useful. If an operator was going to move, something was likely to happen! Even when messages couldn't be decoded, a lot of useful information could be gleaned from 'traffic analysis' – RSS speak for analysing the procedures, preambles, etc.

Before WWII, the RSS (the War Office designation was MI8c) had been set up by the War Office as an autonomous organisation for MI5, because MI5 had no experience of radio. Looking at it from the opposite direction, RSS' main customer was MI5, but when

RSS started decoding the traffic all the foreign intelligence came under MI6. Naturally this caused arguments about the organisations to which the decoded messages should be distributed. Normally, Bletchley Park would give them to MI6 Section 5, the counter intelligence section. The Major in charge of Section 5 was so security-minded that sometimes he wouldn't let half the people who should have seen the messages actually have them, only his own chosen few.

Abwehr station in Hamburg.

In 1941, MI6 Section 8 took over RSS 'lock, stock and barrel' under Richard Gambier-Parry, who by this time was a Brigadier. Gambier-Parry had originally joined SIS to expand their communications operation, because before 1938 the British Government had no direct radio communication with its embassies and agents around the world. In 1938/39 Gambier-Parry was very busy recruiting people to what post-war became the Diplomatic Wireless Service. Known as 'Main Line working', this remained one of the main activities of SCU1 throughout the War. The embassy stations were effectively pirates, because they operated without licences or often the permission of the host country. The radio operators who went out to these countries had to go in civilian clothes and maintain cover, an example of this being that the operator sent to Ireland had to double as a butler.

Box 25

By the end of 1940 RSS had a large and effective VI system, with hundreds of logs flooding into Box 25, Barnet. But RSS were hampered by the fact that, like Pat, most VI's had day jobs.

In May 1941, when SIS took control of RSS, Gambier-Parry appointed one of his deputies, Colonel Maltby, as a director of SCU3. Prime Minister Winston Churchill had long believed in the value of signals intelligence, and through his personal intervention MI6 was given prime control of the RSS operation, still under the designation MI8c. An early result of this decision was to establish full-time intercept stations. At Hanslope Park, near Milton Keynes, the first large RSS special purpose intercept station was soon being planned. RSS started writing to VIs, to ask if they would be interested in joining this special unit. There were all sorts of civilian conditions attached to working for it, so although it was overtly part of the Royal Corps of Signals it was made to sound a bit like a home-from-home, virtually a civilian job in uniform. The salary was quite reasonable and a number of young VIs accepted it keenly.

This is when Pat was invited to join the unit. Nothing happened for some months, but in November 1941 he was asked to return the Hallicrafters receiver, destroy all the paperwork associated with being a VI and report to Arkley View, Barnet. Pat became 2600077 as a recruit to this special Royal Corps of Signals unit. He was one of four radio amateurs who joined RSS that week, the others being Jimmy Adams, GM5LF, Peter Gourlay, GM3LO, and Ron Delahunt, G4QD. They all went into civilian 'digs' in Barnet. Each was given an *Army Book (AB) 64 Part 1* – but not Part 2, the army pay book – a uniform and a slip of paper signed for their Commanding Officer, stating that they were not required to wear it. Even so, they did always wear uniforms while on duty and later found that if they wore it off duty as well, they were warmer and got preferential treatment. The discrimination unit at Barnet was run in a highly effective manner by Major Kenneth Morton Evans, G5KJ, supported by his discrimination staff, almost all of them being ex amateurs. Despite the distraction of air raids in 1940-41, the secret listeners were spending hours searching the HF bands. Whilst at Barnet, Pat passed a very informal and – probably most importantly – unanticipated army medical 'A1'. His temperature was not even taken!

In those days there were three grades of Morse test; 'A' was 25WPM, 'B' was 20WPM and 'C' was 15WPM. The day he reported

The slip of paper that shows Pat did not necessarily have to wear a military uniform.

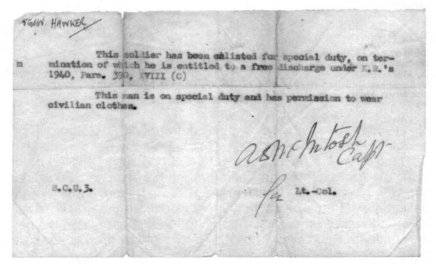

at Box 25, Pat went to nearby Ravenscroft Park, a small RSS intercept station that could conduct the tests, late one afternoon, by
which time all the others had taken their test. Pat sat down and the
examiner asked him what speed he would like to be tested at. "Let's
try 25", replied Pat. He took out a fountain pan and started scrib-
bling away when the Morse was sent. The examiner looked utterly
surprised and commented that he had never seen anyone try to
copy Morse before with a fountain pen, it was always with a pencil.
The sight of the pen must have distracted the examiner a little,
because he rattled away at slightly over 25WPM. Consequently,
when Pat handed over his copy sheet the examiner didn't even
bother to read it properly, he just gave Pat an 'A' pass.

Pat later discovered that hardly any other VI who joined the
service even attempted the A-grade exam, they all settled for 'B' or
'C'. Out of the four, Pat was the only 'A' grade pass, which in a way
turned out to be his undoing. After about ten days, the powers
that be decided he would be better suited as an operator, rather
than working in 'discrim', sorting the logs out. He was duly dispatched to Hanslope Park as an intercept operator, while the other
three would remain in the comparative comfort of Barnet for the
rest of the war. Perhaps opting for the 'A' grade Morse exam was a
tactical error on Pat's part, but it led to him having a far more interesting war than the others.

Hanslope Park

At the time of Pat's arrival at 'The Farmyard', the only intercept
station at Hanslope Park was in a small, hastily converted granary.
There were about six HRO receivers. Pat recalls the long slog by
foot back from one of the four nearby pubs and one tea shop.

The village of Hanslope served mainly as a dormitory for
those employed in the railway and printing works of nearby
Wolverton, and the Hanslope Park Estate had been the scene of
a murder before WWI, when the owner was shot dead by his
gamekeeper. The local vicar also made a name for himself in *The
News of the World*, for alleged activities on visits to wicked London. Nevertheless, 'The Farmyard' wasn't without distinction.
Amongst those who worked there for a time was Alan Turing, a
brilliant mathematician who became the pioneer of digital com-
puting mathematics and advanced cryptography. Turing himself
wasn't exactly popular with the local constabulary, because he
used to wear a gas mask while riding his bicycle to work. In fact
this had been a sensible precaution, because they were working
in a rural area and he suffered from Hay Fever. Pat doesn't actually recall meeting him though.

A Bit of Controversy

After about six weeks they moved the intercept station and set it up in The Lodge, which was a few hundred yards away from Hanslope Park itself, on the road towards Wolverton. The initial operators all lived in The Lodge, while new recruits who were arriving at the rate of two or three a week all had to live in Hanslope Park. Some of those who had to live in the Park took to expressing their dissatisfaction in verse. With feeling, George Proctor, GM8SQ, penned the following:

> *When first I came to Hanslope*
> *And saw its 'lovely' huts*
> *I said the Army's lousy*
> *Why did I join? I'm nuts.*
>
> *But now I am confounded*
> *For it is plain to see*
> *If I think the Army's lousy*
> *It thinks the same of me!*

Conditions at The Lodge were more relaxed than at the Park, with the operators pretty much left alone to get on with their work. There was a requirement for a 24-hour listening watch so the operators themselves worked out various systems of covering it. The worst was a spilt shift that began at 2:00am, followed by a break, then more listening around lunchtime. In May 1942, when the new and much larger station opened, all the operators moved there and changed to a fairly conventional 8-hour shift pattern. At The Lodge, Pat shared a room with WW (Bill) Peat (who would become GM3AVA), Johnny Bowers, G4NY, and Des Downing, GI3ZX.

2600077 Hawker JP, in 1942.

The Hanslope Park accommodation was 'interesting'. Four new brick-built huts had been erected for the operators. It was all part of their conditions of service. There were individual sleeping quarters, with cupboards etc. When new operators got there though they found that although there were 'rooms', there were no internal doors. Also, bunk beds had been put in each room, so people had to share. Blankets were provided, but no sheets. To add insult to injury, there were no flush toilets either.

The day Pat arrived they went into Hut 4. There were already a couple of VIs in there, but also a lot of 'General Duties' personnel – people who were medically unfit for anything other than undemanding manual work. Pat described some of these "old lags" as giving him quite a culture shock. During the first night there was a lot of noise in the hut, so the next day Jack Kelshall, G4FM, went along and complained to the Adjutant that the VIs couldn't sleep, so needed to have their own accommodation. All the General Duties staff were duly moved out and Hut 4 then filled up with more operators.

As an Intercept Operator, each bay had two HRO receivers in it. There were 32 bays of double receivers and two bays with three, these being used for general search purposes. Operators would be brought a schedule card, so that they would know what they were supposed to be listening for. There were no names, just codes like '2/520', but the great circle bearing of the station would also be provided. The intention was that the operator would go to a distribution panel and select the appropriate antenna for the bearing that was to be listened to. The antennas would have been rhombics or Vee beams, about 70ft above ground level. Outside the transmission lines were all open wires, but inside they were connected to baluns, so the distribution panel used coax… and here was where Pat came across coaxial cable for the first time.

Engineering at Hanslope Park was under the control of Major Dick Keen, whose book *Wireless Direction Finding* was for a long time the classic text in its field.

All the masts at Hanslope had all been erected by 'Digger' Buick, G3XJ, and his gang of Non Combatant Corps. Digger was Australian, sported a non-Parliamentary vocabulary and had been recruited from EMI by another very well know radio amateur, Dud Charman, G6CJ. Mainly conscientious objectors, Pat remembers the NCC men as being intellectual and studious people. Having been previously engaged in the clearance of bomb sites in London, Pat remembers that they worked like mad to put up aerial supports at Hanslope, Whaddon and Forfar. Maybe it was Digger's constant stream of obscenities that spurred them on.

With its plug-in coil packs, the National HRO receiver was an advanced model of its day.

Hanslope Park had quite a selection of directional aerials. They beamed in all directions, no more towards Europe than anywhere else. As an operator, after selecting the one you wanted you would return to your desk to listen, but it was not unknown for someone else to then come along, take your antenna for themselves and plug you into one that was beaming in completely the wrong direction.

As well as being a 'group leader', Dud Charman also spent time at Hanslope Park. He produced the low-noise signal distribution amplifiers, which had been remarkable at the time. There's a story associated with these amplifiers in that they had originated as wideband amplifiers for television distribution at Radiolympia exhibitions before WWII. 807s were used at low power, which enabled the amplifiers to operate on the straight part of the curve and be extremely linear. Generating little heat also meant they were extremely reliable. Banks of them were used at Hanslope Park and remained in service long after the war, when it became part of the Diplomatic Wireless Service. A Canadian professor who was there to advise on cryptography and engineering matters was so impressed that he told Dud if he wanted a job after the war to come and see him in Canada!

Post war Dud became famous for his 'Aerial Circus', in which he would demonstrate visually the radiation pattern of aerials with miniature ones running at around 3GHz. A video recording of the circus can be borrowed from the Society.

Pat was a great enthusiast of the rhombic aerial but, as he puts it, "you were lucky if you could get the right rhombic, but sometimes you did. They were very good aerials". By mid 1942 Hanslope Park was an efficient and well-run station, under the charge of Captain Prickett. He had been a merchant navy operator, but had subsequently worked in South America. Upon returning to Britain he had joined RSS. Pat remembers him as being a very pleasant, easy-going officer. On one occasion Pat remembers asking him if he should ask about doing something, or just get on with it. "Go ahead", said Prickett, adding "Don't ask me; I might have to say 'no'." Pat learned later that Prickett had run into difficulties for putting his precepts into practice, because he was eventually Court Martialled.

Throughout the time Pat was employed by the SCUs, he never spent a day in basic training. At Barnet they route marched people around, but at Hanslope a Captain Ash thought that he should make real soldiers of the men. They brought in a Sergeant Major to make this happen. One day the men were route marched around seven miles of countryside, then put on duty in The Lodge. The Interceptors didn't like this one little bit and reacted against it by entering in their logs that they were too tired to listen. All the logs were forwarded to Barnet, where they caused a fuss. It ended

up with the Captain being told to go easy, although they never entirely stopped his attempts to militarise the unit.

Conditions gradually improved at Hanslope Park, with a medical hut, a large mess hut and a NAAFI.

There was a weekly Pay Parade at Hanslope Park. All the General Duties people would have an AB64 Part 2, which was the army pay book. The RSS people only had AB64 Part 1. "You would be handed over a little envelope, containing your wages", remembers Pat. After moving to Weald as SCU1 in the spring of 1943, by which time he had risen to the giddy heights of being a Lance Corporal, none of the officers were ever required to attend a pay parade.

Other Stations

Weald was a part of a two-way communication facility. The receivers were at Weald, while the transmitters were at Calverton, not far from Whaddon. The operators were a mixture of civilians (wearing Royal Observer Corps uniforms) and National Service and Special Enlistment Royal Signals personnel. Pat and the other operators were still accommodated in Hanslope Park, but were moved to Hut 19, which was allotted to those who had transferred to Weald. Bicycles would be used extensively, as they brought the local cinemas at Wolverton, Stony Stratford and Newport Pagnell within reach. Weald and Tattenhoe are not far from Bletchley Park, where Pat would sometimes go because they had an Amateur Dramatic Society. Not having been bitten by the acting bug whilst at school Pat didn't get involved on stage, but

A Bit of Controversy

recalls that the shows were worth watching.

In early 1943, Lord Sandhurst, who had by now left RSS and was running Section VIII(P) for SCU1 at Whaddon Hall, went to Hanslope Park to look for more operators for his control station at Weald. About 20 of the Hanslope operators volunteered to transfer, including Pat's good friends Johnny Bowers, 'Bill' Peat and Reg Cole. As a Group Leader, Pat had more difficulty in transferring, but he succeeded in doing so within a week or two.

Proving they did sometimes wear civilian clothes, Frank Watts, G5BM, Watson (Bill) Peat, GM3AVA, and Pat, at Gayhurst, near Hanslope in the summer of 1943. They used to swim in the river.

If Hanslope Park had been a culture shock for Pat, Weald was a technical shock. Until Digger and his crew arrived in 1944 and improved the set-up with tall masts, Weald's aerials were crude, semi vertical wires suspected from relatively low cantilever wires.

For a while, as a sideline, SIS maintained radio contact from Weald with ships on secret missions. In a way it was quite exciting work, but because it was a continuous watch while the ships were at sea, there were also long periods of boredom because radio silence was usually maintained. Voyages could last three or four days. When the ships did transmit they would send a Syko message. Weald had a Syko machine, to decode it. A typical message might go along the lines of; 'Mission success. Three bodies and four sacks of mail.' The ships would dock in South Devon or Cornwall, but eventually the base was moved to the Scilly Isles and Section VIII set up their own Mk.10 SIS transmitter. Consequently Weald lost that bit of work, but they weren't at all sorry about it because the continuous watch element wasn't something that anyone welcomed. One of the operators favoured for the long watches had been a good operator, but had developed 'brass arm' from his merchant navy days. These days we call it repetitive strain injury. Pat remembers that there was a cat at the station that would sit by the Morse keys. Sometimes it would walk across them too, which made for some interesting keying.

One of the clandestine stations that Weald maintained a short daily contact with was the Beagle weather group. Led by Albert Toussaint, it remained on the air until the final liberation of Belgium in September 1944.

At that time there was no mains electricity in Weald; everything ran from batteries. A generator man used to go out to charge the batteries from a diesel generator. After Pat left it became such a busy station that mains electricity had to be installed. Prior to mains electricity, the men were told by 'Sandy' that they didn't have it because it made too much noise and running from batteries was much quieter. It is more likely that they couldn't get mains installed early on because people were thinking about other things. Pat feels the agent traffic wasn't treated with as much respect as RSS intercepts. Also, the conditions at Weald could be

decidedly unsavoury, with a bench that served as a bed and blankets that were shared but never washed. In December 1943 accommodation for the operators moved to the new Section VIII hostel at Tattenhoe, just outside Bletchley.

Section VIII under Richard Gambier-Parry moved originally to Bletchley Park, where Station X was located, but then it was decided to set up a separate facility at Whaddon Hall, although Bletchley Park would remain the decoding centre for their agent traffic. Their main purpose then became distributing the Ultra decodes from the Y service, which would then be sent to Whaddon Hall where they would be re-encoded differently and then sent out to army commands from Windy Ridge – which was a communications facility staffed largely by army personnel, although some got special pay. Despite the fact that it is located very close to Whaddon, Windy Ridge worked as a separate unit, indeed Pat never even saw the place close up.

The Main Line working to the embassies moved out of Whaddon Hall itself into large huts that stood in front of it. They worked on a completely different basis from Weald and was a service that after the war Pat himself would eventually end up working for.

In about July 1946 Hanslope Park was reconfigured and took over the Main Line working, plus the Whaddon engineering labo-

Weald SCU1 (Section VIIIP) control station working to secret agents in France, Belgium and Holland. Also to the "Liberation" SCU9 stations in Belgium and Holland after the liberation of France in 1944. Mainly HRO receivers, but also AR88 and AR60. The associated transmitters were at nearby Calverton.

ratories and workshops. In 1947 it became known as the Diplomatic Wireless Service.

Operation Sussex

In May 1944 Captain Tricker became Major Tricker and set up a
new unit – SCU9. Taking-in several staff from the Weald and Nash
stations – including Pat and several of his colleagues – SCU9 worked
to agents in Norway, Denmark and Holland, whereas Weald
worked mainly to France and to a lesser extent Belgium. SCU9
was a small, mobile unit, destined to travel to Normandy after D
Day. It was divided into an advanced party with three operators
and one small signals vehicle, plus a main unit of nine operators.

Attached to No.2 Intelligence (Underground) Section, the unit
was partly for use in connection with WWII's largest single Allied
Intelligence operation – 'Sussex'. This involved parachuting 50 two-
man teams of French secret agents in a wide sweep from Brittany
to Belgium. Equipped with radios, their function was to report German troop movements and other intelligence, independently of
existing intelligence or resistance groups. Operation Sussex was
run jointly by MI6/OSS/BCRA. The responsibility for radio links
was divided between Section VIII(P) (Brissex) – who would work to
the teams in the British zone – and OSS-Special Intelligence (Ossex)
– who would work to teams in the American zone.

The agents were trained at TS7 (Training School 7) in St Albans,
which was established specifically for the operation in a fine, old,
ivy-covered mansion called Glenalmond, surrounded by park-lands
and with its own ornamental lake. Operations were under the con-

These young
Belgians provided
hundreds of daily
weather reports
for the RAF to
Section VIII's
Weald control
s t a t i o n .
Organised by
Albert Toussaint
(Beagle) - left - the
group operated
until the liberation
of Brussels in
September 1944.
Not all survived.

trol of Colonel Malcolm Henderson. Even by the standards of SIS, his qualifications were unusual, as he had served in the French Foreign Legion and been an opera singer. In fact many of the instructors at Glenalmond had odd qualifications; one had been an acrobat in the Balkans, while another was an Oxford Don. In the mornings students would arrive for lessons, while in the evenings they would go down the pub. In the mess hall, wine was

available – a highly unusual practice in the UK, even for those on special duties. One of the trainees was only 17 years old, while many were reservists who had military experience in France or North Africa. Exercises included the identification of German ranks and insignia, cover stories, how to avoid being shadowed or captured, how to face interrogation, plus the use of radios and special codes.

Those intended as radio operators were given additional training at the Section VIII wireless school at Hans Place, which is behind Harrods in Knightsbridge, and during practice exercises local to St Albans and also further afield. They also learned simple repairs and were given briefings from agents who had recently returned from France. With a signal plan entitled 'Wayfarer', the first agents were dispatched in April 1944. The 25 Ossex teams were to work to the OSS Special Intelligence station 'Victor', located at Hurley in Berkshire. They were equipped with the Whaddon-built MkVII 'Paraset' agent sets and 'Ascension' 35MHz FM radio telephone equipment

Main party of the 18-strong Special Communications Unit No.9 (SCU9) with 12 Section VIIIP operators and driver/ mechanic, driver and two despatch riders in 1944, just prior to leaving for Normandy to provide a link between SIS and the main Army Command. Pat is the tall one in the back row.

Photo: Dick Rollema, PA0SE.

The Whaddon-built MkVII 'Paraset'.

A Bit of Controversy

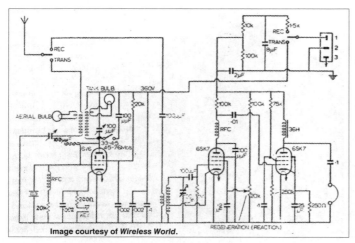

Image courtesy of *Wireless World*.

Circuit diagram of the 'Paraset'.

for working to aircraft.

Early on in the planning, the British revealed to the Americans the existence of ground-to-aircraft agent radios. The Americans dubbed this 'Klaxon' communications. The Americans didn't have equivalent models of their own at the time (their 'Joan-Eleanor' equipment was not available until late 1944), so SIS supplied Ascension equipment to the Ossex teams as well. As WWII progressed, telephony was becoming an increasingly necessary means of communications, because it was difficult to find proficient Morse operators capable of working behind enemy lines.

One of the problems sending secret agents to France was getting them to blend in. They needed to wear exactly the right type of clothing and carry exactly the right kind of things in their pockets, but all of this was in extremely short supply in Britain. The best that could be done was to send them to France in well tailored suits. The wise ones changed into soiled garments as soon as they arrived. Some of the teams had great success, while a few were sadly eliminated.

On 23 March 1944 General Eisenhower took over all control of all the secret service activity connected with Operation Overlord, the D-Day invasion of Normandy. Liaison between SIS and the 21st Army Group became the responsibility of 2I(U).

Mk3 Sten gun.

Before going to Normandy, it was required that Pat fire a certain number of shots. This was so that he could be given a Lee-Enfield rifle and later a Sten gun, which was a small, light, sub machine gun. He remembers firing the rifle, and he also remembers Johnny Bowers doing so and coming back, having failed to hit quite a large target. "We were not good soldiers. We were Fred Karno's Army absolutely," says Pat, adding "We still looked upon ourselves as civilians in uniform."

Travel

The whole of the D-Day invasion of Normandy was delayed by bad weather, then parts of one of the concrete landing-piers sank, so the only Mulberry Harbour built was at Arromanches. By the time he set foot on French soil it was D+50, Pat having spent five days on board an American LST (Landing Ship (Tank)), three of them anchored off Southend Pier. Pat remembers that the American ships didn't seem to have any experienced sailors on them, but they did have decent food.

After landing in Normandy, SCU9 established a base at St Gabriels, although they soon fell foul of the locals for using the convent as a mess hall, even though it had been so used by the Germans before their arrival. There were air raids by the Germans, but Pat feels they were more in danger of being struck by ack-ack shrapnel than a bomb. During raids some of the unit would

St Gabriels, where 2I(U)/SCU9 made themselves unpopular with the locals.

reach for their steel helmets, but it wasn't their heads they chose to protect with them. From St Gabriels a number of later Brissex agents were infiltrated across the lines, some with MkXXI battery operated radios concealed in the saddle of donkey.

While in Normandy, Pat met a sailor who had manned motor torpedo boats. He was complaining bitterly that the day the Channel was declared clear the Navy stopped paying danger money, which didn't go down at all well. Of course SIS were running agents across to Confrérie de Notre Dame. This was a catholic intelligence movement, based in Paris and headed by Gilbert Renault-Roulier. Known as 'Colonel Remy'. He was one of General De Gaulle's most trusted agents and was connected with the French film industry. He ran a big network, including radio operators, but some of his people got 'turned' in the end. He organised a system via Breton fishermen, sailing out of Brittany (who were overlooked by the Germans). They would make contact with an SIS boat that was disguised as another French fishing vessel, take agents out and bring back mail and questionnaires, because it was too dangerous to use radio to pass long and detailed messages.

SCU9 moved from St Gabriel to near Port en Bessin in the

A Bit of Controversy

Watson 'Bill' Peat, operating a Mk3 transmitter. The power supply was in a separate box that was heavier than the transmitter. It had a tendency to catch fire. Pat's own typewriter is in shot, which he later used to type up 'TT'.

American Zone, until the Americans told them to leave, then on to the grounds of a chateau near Juaye, south of the city of Bayeux. SCU9 was attached to No.2 Intelligence (Underground) Section, which included a number of Intelligence Corps cipher clerks who had recently joined-up and had been given six weeks training at Bletchley Park. Pat was with them in Normandy until about 24 August 1944. By that time Bill Peat, who later became GM3AVA, was attached to an officer and sent to go out and recover agents who had been overrun. After that, the idea of attaching communication operators to individual intelligence operators began to take off.

Paris has fallen – or has it?

When the BBC, quoting Radio Algiers, announced on Tuesday 22 August "Paris has fallen", Eskimo Nell, which was what SCU9 called their 15cwt signals vehicle, was dispatched with CQMS Gerry Garrish, Bill Peat and 2I(U) officers towards the French capital. In fact the BBC had been premature with their announcement, because by the very next day they were announcing that the Germans were not only still in control of Paris, but looked like crushing the revolt.

The Paris police, who had up to that time been collaborating with the Germans had risen up against the occupiers. SCU9 had their minds on another matter though. Peat and Garrish hadn't made contact with the main unit at Juaye. Moreover, there had arrived at Juaye an urgent personal message for General Charles de Gaulle. The message was causing a great deal of embarrassment at Juaye, Weald and Broadway Buildings – the headquarters of SIS – because it couldn't be delivered. After several days it was decided that they needed to discover what had become of the advance party.

On the morning of Sunday 27 August Pat was hurriedly dispatched by Jeep to Paris, along with two intelligence officers and a driver, the intention being to dash into the city, discover what had happened and deliver the message by hand. By that time Paris had indeed fallen (it actually fell on the evening of Friday 25 August, and celebrations were held on the Saturday). Peat and Garrish had reached the city and on the Saturday evening had set themselves up in an abandoned Gestapo building in Avenue Victor Hugo. They didn't attempted to make contact though, because there were so many celebrations taking place. On the following day they moved out to Villa Montmorency, which was a closed estate of high-class buildings. The unit had authority to take over any buildings that had previously been requisitioned by the

A triumphant Charles de Gaulle walking along the Champs-Élysées shortly after the liberation of Paris.

Germans, but they were not allowed to requisition any building that had not been affected during the occupation. Pat arrived in the afternoon, making immediately for de Gaulle's HQ in the Hotel de Ville. Pat sat outside in the Jeep, while the message was delivered to the General inside the Hotel de Ville. They then drove to Villa Montmorency, where by this time the small 2I(U) party had set themselves up in one house, while SCU9 in the form of Garrish and Peat occupied another on the opposite side of the road. Pat joined in operating the radio.

A day or two later Major Tricker arrived, along with two more operators, Roy Wilkins and Johnny Bowers. Pat and his colleagues stayed there for about six weeks. In the end they were also handling traffic for the British Embassy. For this they were isued with a special piece of paperwork, signed by the French Head of Security, to confirm that they were part of the Anglo-French Liaison.

Paris Experiences

Villa Montmorency was quite a way from central Paris, so it was a long walk to the Champs-Elysées (rather less to the Eiffel Tower).

Pat was lucky enough to climb the Tower on the first day that it reopened. Unfortunately the lifts were not working, so it was a case of trudging up the stairs. While on his way up, a young Frenchman came rushing down and brushed past Pat. He seemed to be in a real hurry. Pat got to the first landing and had a look around, but went no higher. Later he discovered that De Gaulle had been making a speech at the Trocadero, near the Eiffel Tower. Someone had tried to shoot him from the Eiffel Tower and Pat is pretty sure that the gunman was the very man who had rushed past him on the stairs.

After the Paris Metro had reopened, one day Pat wanted to go into central Paris with Johnny Bowers, so they joined the queue to buy tickets. The French in the queue created a fuss at this. They simply wouldn't allow Pat and Johnny to pay, insisting instead that their liberators travel free. Apparently the German military had never been required to pay to travel on the Metro, so the French insisted that the British should enjoy free rides as well.

One day Pat and Johnny Bowers were walking along the Champs-Elysées, when a huge procession started to pass them by. It was a mixed parade of American troops, marching fifteen or more abreast, interspersed with French military bands, but no British of French troops. Naturally enough they thought it must be a victory parade. Now it was always claimed by the French that it was their own troops and the internal resistance that had liberated Paris. Apparently the parade really enraged the civilians, to see no French troops, but they hailed Pat and Johnny and insisted on buying them drinks in a bar. *Vive l'entente cordiale!* When reports of the parade got into the British press it created even more of a stink, because there had been no British troops in it. The War Office duly had to issue a press release, saying that there had in fact not been a victory parade at all! They put a spin on it, saying that it was just a few American troops on their way to the front line. De Gaulle had been at this parade and had walked off the stage in a huff when he saw that no French troops were part of it. The whole thing taught Pat one thing… never believe an official communiqué!

By 10 October 1944 Main Line operators had arrived for the British Embassy and the SCU9 operations in Paris came to an end.

Brussels

The main unit from 2I(U) (No.2 Intelligence Underground section) and the main part of SCU9 had stayed in Normandy until the liberation of Belgium in early September 1944. This took place at amazing speed. The Allies rushed across and reoccupied it in hardly any time at all, setting up a station there with 2I(U). At the time they were handling SIS traffic for the 21st Army Group, but 2I(U) section and SCU9 always remained physically separated from 21AG (HQ).

A V-bomb subsequently landed on the building next door, largely destroying the building and killing four people in the house that was hit. The SCU unit had to relocate, but Brussels remained their HQ until the end of the war in Europe.

The Netherlands

From Paris, Bill Peat had gone on with an intelligence officer up into Eindhoven, as part of the Arnhem landings. They arrived in Eindhoven the day it was liberated. Bill Peat stayed in Eindhoven and the other operator, Stewart ('Chick') Francis, who later became G3AVI, went on with another intelligence officer across the Nijmegan bridge into the area between the two branches of the Rhine. He stayed there a couple of days and set up a station to work back to Brussels or Eindhoven, until he looked out of the window and noticed a distinct lack of British troops around them. Upon investigation they discovered that all the British troops had fallen-back beyond the river at Nijmegen, so Francis hurriedly withdrew too.

Chick Francis went to Eindhoven and joined Bill Peat, staying there for several months. It became an extraordinarily busy station, because the Dutch Intelligence had set up HQ in the Abbe Museum, which is an art gallery in Eindhoven. The Dutch had their own network and passed all the relevant traffic to 2I(U). Peat and Francis relayed it back to Brussels and the UK. They became one of the busiest sections of SCU9.

Along with an intelligence officer, Pat went from Brussels to Nijmegen at the end of October. Pat set up a station in a Dutch home on the outskirts, a standard station for SCU9 being a National HRO receiver and a Mk3 transmitter, all powered by a petrol generator. Incidentally, the Mk3 transmitter with its 807 PA stage is said to have been based on designs in amateur radio handbooks. The very next day another group arrived. This was IS9 (Intelligence School No.9), under Major Hugh Fraser. In fact it was a section of MI9, the escape and evader unit that organised the escape of PoWs and pilots whose planes had been brought down over Belgium or France. Over the course of WWII, the evader section was responsible for 2,000-3,000 RAF pilots making it back to Britain. Most of this was via the long route down through Spain, to Gibraltar, albeit at the cost of around 2,000 French Belgian and Dutch civilian helpers.

A Bit of Controversy

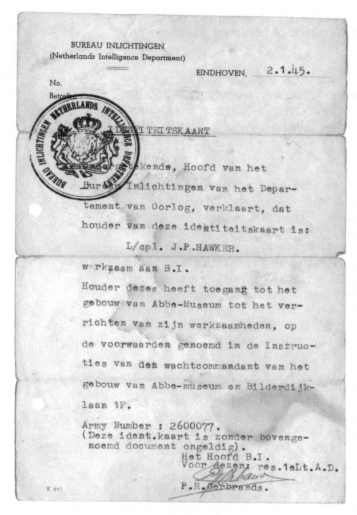

The paperwork that confirms Pat was authorised to work for the Dutch Internal Radio Service.

MI9 had linked up with MI6 in 1940. MI6 offered to help with the setting-up of an escape route out of France and set up the Pat Line (named after Pat O'Leary). This was later followed by the Comet Line to get people out of Belgium, but the Germans were always trying to infiltrate the escape lines. It got to the point where they would plant English-speaking people in RAF uniforms in various places, pretending to be parachutists trying to get back home. Pat discovered just a year ago that the father of a neighbour had been the Colonel representing MI6 at MI9 and was an author of a number of books about the service. His mother had been a member of the Comet Line.

In December, in the Battle of the Bulge, the Dutch discovered that the intention of the Germans was to sweep across Holland and retake Antwerp, which by that time had become the main sea port for the allied forces in northern Europe. The result was that the Dutch clandestine network hurriedly borrowed the Eskimo Nell vehicle from Bill Peat and put their own equipment into it, so they could carry on their own clandestine network. But it never got used.

During Christmas in Brussels, Major Harry Tricker asked Pat if he would be prepared to work for the Dutch. They badly needed some experienced radio operators. Tricker, a former Chief Petty Officer, had been recruited before the war from Flowerdean, which was a Y service station. Although a senior rank, Pat doesn't remember him ever giving anyone a definite order. The way it worked was that they asked people to do things, rather than ordered them. As Pats succinctly puts it, "Section VIII was very civilised!"

Pat went back to Eindhoven with SCU9 Sergeant Bert Lawler, who was a Post Office man. There was one station at the museum, and the original one in a private house. Bert went to the house and Pat - soon as a Local Sergeant - went to the museum. For a short while, when the network was told not to go on the air at all, Pat went to a semi-underground water tower south of Eindhoven, but basically he worked at the museum until the end of the war, gaining much respect for the Dutch secret radio operators.

Pat retains enormous respect for the Dutch operators - amateurs and professional radio-telegraphists - who suffered grievous losses. It is likely that more Dutch than British pre-war amateurs list their lives in WWII.

Christmas 1944. Pat in Holland.

Double Agents

Generally, RSS interception worked well, particularly when it came to double agents. The Germans would send agents over to Britain with a radio transmitter. RSS knew they were coming, so they would immediately be picked-up by MI5 officers, taken to Latchmere House on Ham Common, Richmond, and pressurised into becoming double agents. Some whose arrest had been reported were tried and found guilty under the Treachery Act, and were hanged. Most co-operated. In some cases double agents were then given radios, to send misleading information back to Germany. One of the main operators who oversaw the double agents was Ronnie Reed, G2RX, who had been a VI and was seconded from RSS into MI5. Over the course of the war Ronnie collected all sorts of equipment, including a number of German items. One

The Iron Cross. It was awarded to men and women of all ranks within any branch of the Wehrmacht, Waffen-SS or German auxiliary service organisations.

was an adapter that could be plugged into a radiogram, to convert it into a transmitter. He also had an Iron Cross (Second Class) that had been awarded to Tate, who was one of the agents that had been 'turned' and worked for the British. Tate continued to transmit to Germany throughout the war and was eventually awarded the Iron Cross (First Class), so he passed his Second Class Cross on to Ronnie, saying he had been as responsible as himself in getting it! Today, Latchmere House is part of the British prison service.

Pat himself never had contact with the double

agents. RSS people knew they existed, but they were controlled by the so-called Double Cross Committee. This name was derived from the Roman Numerals 'XX', because the committee was actually Committee No.20. "A very suitable name", as Pat puts it. On a small scale the system had run from the beginning of the war. Pat learned after the war that MI5 had eventually gone through all the Abwehr documents, to see how much they had cottoned onto their agents being knobbled. There was a report from an Abwehr officer who had been told to examine all the reports from their agents in Britain. He was convinced their agents were being controlled by the British and duly reported the fact to his superior officer. His boss bounced the report back to him with a message along the lines of 'Do you want us to be sent to the Russian Front?' The report was duly filed away and no action taken. The Double Cross operation continued until the very end of the war.

It wasn't just the British who ran double agents, the Germans did it as well, and they did it very successfully. In fact they ran the entire SOE operation in Holland for 18 months. This led to the loss of at least 47 Dutchmen who were sent over as secret agents, ten RAF planes on special missions and the delivery of thousands of arms into enemy hands.

Falling out

The RAF wanted SOE closed down as a result of the Germans having infiltrated the Dutch network, and they nearly got their way. They were only saved by the intervention of Winston Churchill and Clement Attlee. Churchill, who had been responsible for the setting-up of SOE, took away SIS's sabotage and propaganda arm and gave them to SOE, resulting in friction between SIS and SOE. SOE also fell out with the RAF.

Any rivalries that may have existed between members of the Royal Navy, the Army and RAF pale into insignificance beside those that were endemic within the various secret services, and not just on the side of the Allies. There was, for example, absolutely no love lost between the Abwehr and the SD (Gestapo). In 1944 Winston Churchill was moved to write: "The warfare between SOE (Special Operations Executive) and SIS (Special Intelligence Service) is a lamentable but perhaps inevitable feature of our affairs". One reason was that SIS felt sabotage and covert paramilitary operations tended to stir up German countermeasures, the problem being that intelligence and escape networks prefer to work alongside apparently dormant populations, rather than in an atmosphere of police raids and hostage taking.

By the end of the war the British and Americans had fallen out as well. Montgomery was angry that he had been overruled, which is partly why Operation Market Garden had gone wrong at Arnhem. His great plan had been to go up through Holland, leading a joint British-American operation into the heart of Germany. Eisenhower, who was the Supreme Commander, Allied Forces, had wanted a broad front into Germany from the east generally, and to combine that with an advance through Holland. Operation Market Garden was delayed while they disputed the issue. Although Eisenhower finally gave way, the main American forces remained committed to a broad front attack on Germany. Arnhem proved the 'Bridge Too Far'.

Although the signals directorate of SOE and SIS's Special Communications section included significant numbers of former radio amateurs, there was little co-operation between the two and from 1942 both tended to go their own way. Both developed radio links with resistance groups from high-flying aircraft, but using different equipment.

The last German air raid

Working on the assumption that the RAF would be sleeping-off their New Year celebrations, on 1 January 1945 the Germans mounted their last major air raid on Allied airfields. Pat was walking across a bridge at the time, on his way to the Abbe Museum, but was uninjured. His feelings about the sounds of war are that artillery fire was quite comforting, because it meant the fighting was a long way away. Small arms fire was more worrying, because it meant the fighting was close. Apart from when crossing the English Channel, he never really felt himself in much danger. It didn't compare to the dangers faced by the fighting forces or to the dangers faced by the radio agents – many of whom were young, hastily trained and then infiltrated behind enemy lines by being dropped by parachute, infiltrated from Spain or landed from a disguised fishing boat – it was nothing. Many of them didn't live to see the end of hostilities, which evoke feelings of sadness to this day. An agent landing in occupied Europe could have been controlled from the outset, the resistance network having already been infiltrated, betrayed, captured, injured or 'turned'. The odds were against them remaining at liberty and being successful, although many were.

June 1945, Les Fragle - another SCU operator - and Pat.

Pat describes his war as having been more "glamorous" than

A Bit of Controversy

Above: The residence on the Rhine with the spiral staircase.

Right: A coffin set like the one that had to be dragged up the stairs.

Bottom: The house in the Rhineland where Pat spent October 1945 to January 1946.

"glorious", adding "Looking back on it, the army, navy and air force had a dislike for intelligence anyway… and they were probably justified. Nevertheless, Signals Intelligence (SIGINT) and human intelligence (HUMINT) played an impoortant and significant role in reducing the length of the war."

In occupied Germany

When the war ended, Pat stayed in Holland for about three weeks. Much to his disappointment he was then told to wrap-up Eindhoven and move into Germany.

It was while Pat was in Germany that he picked up a gun for the first and only time in earnest. At that time there were a lot of displaced persons in Germany, many from Eastern Europe. They wanted desperately to get back to their homelands, but couldn't, so started to raid German farmhouses, robbing and stealing what they could. Such an incident took place one night near where Pat was at one time based. He rushed out, carrying a Sten gun, and accompanied other intelligence 'soldiers' to where the action was. It is probably fortunate for all concerned that by the time they arrived it was all over. He was there for some weeks, although during this time he also went to Brussels.

In mid 1945 the SCU9 base moved to Bad Salzuflen, a small town between Dortmund and Hannover – then in the British Zone. About ten miles away is Bad Oeynhausen, which was the main military HQ at the time. There was a lot of liaison between groups there. It was entirely an SCU9 operation at Bad Salzuflen, in an old hotel/spa. Pat recalls gleefully that he didn't have to stay at Bad Salzuflen for very long. By that time 2I(U) had become 5CCU (Civilian Control Unit) and Pat went down to Bad Godesburg as an operator with a small intelligence section from 2I(U). First they moved into a little hotel in the town, then a swish place on the banks of the Rhine. In this house there was a spiral staircase that led up to a room on the first floor. This became the radio room. Pat remembers dragging up the stairs a so-called 'coffin set', which was an HRO receiver, and HRO power supply unit, a Mk3 transmitter and a very heavy Mk3 power supply, all built into a wooden box. It weighed a ton!

Later, after UK leave, Pat went back to another (7CCU) house near Bonn, staying there until the end of 1945, taking an active part in the intelligence operations, icluding the cyphering etc.

Still a local sergeant, Pat was demobbed on 14 September 1946, after serving his country for one day under five years. He was discharged under Kings Regulations (KRs), 1940 Para 390/XVIII/a, "his services being no longer required for the purpose for which he enlisted". This KR is normally used for getting rid of undesirables, but Pat's glowing testimonial tends to indicate that it was not the case this time. It meant that he was never on the Z reserve, many of whom were recalled during the Suez crisis in 1956.

Pat's memories of his time in service are at slight odds with the testimonial he was given upon leaving!

Serial no. 5872

DISCHARGE CERTIFICATE.

Army Form B108J

(If this CERTIFICATE is lost no duplicate can be obtained.)

Army Number 2600077

SURNAME Hawker

CHRISTIAN NAMES John Patrick

Effective Date of Discharge 10 November 1946.

Corps from which Discharged **R. SIGNALS.**

Service with the Colours : Years four Days 364

Service on Class W(T) Reserve : Years — Days —

Total Service : Years four Days 364

Rank on Discharge Lce Sergeant (L Corporal).

Cause of Discharge Discharged under KRS 1940 PARA 390/ XVIII/a "His services being no longer required for the purpose for which he enlisted.

Campaigns and Service Abroad N.W.E France 21.7.44 To 17.2.46.

Medals 1939/45 STAR. France & Germany Star.

Military Conduct Exemplary

Signature and Rank **R. SIGNALS.**
Officer i/c Records.

Date 6 SEP 1946 19 Place READING.

ARMY RECORD OFFICE

1 4 SEP 1946

ROYAL CORPS OF SIGNALS
READING.

(S.9728) Wt.13173/184 32,000 R 49 H/y Gp.669

Pat's discharge certificate.

Worthy of mention

Pat remembers a number of amateurs who made a significant contribution to the war effort.

In 1933 Douglas Walters, G5CV, who was the radio correspondent of *The Daily Herald* and George Jessop, G6JP, had conducted pioneering experiments on 56MHz, in two chartered Dragon Moth aircraft. They had established two-way communication at VHF between aircraft in flight and between aircraft and the ground, and it created a lot of publicity at the time. It is largely thanks to these experiments that when WWII began the British had lightweight VHF transmitter-receivers in their fighter aircraft – an essential requirement for the effective use of the first early warning radar chains.

The Royal Naval Volunteer (Wireless) Reserve had been formed in 1933 and the RAF Civilian Wireless Reserve in 1938 – indeed Pat was present at the 1938 RSGB Convention where the CWR was launched. They were both mobilised as war threatened. The first draft of Civilian Wireless Reservists, known as the 'Early Birds', reached France on 5 September 1939. Their job was to form a 'wireless intelligence screen'. They paved the way for the close relationship between radio amateurs and the Y service – the signal intelligence services that closely monitored the radio traffic of the enemy and contributed directly to the code-breaking groups at Bletchley Park. That first draft included Roy Stevens, G2BVN, and W H (Bert) Allen, G2UJ. Roy Stevens

Group Leaders and staff of the Radio Security Service, 1945, which included many leading radio amateurs. In the centre, with the pipe, is Lord Sandhurst, who recruited the amateur VIs in 1939-41.

Pat at Tattenhoe Hostel in 1946.

41

would go on to be the President of RSGB in 1966.

The CWR also brought into RAF Signals Intelligence an enthusiastic radio amateur, Rowley Scott-Farnie, G5FI, who later become head of RAF Air Intelligence in the Middle East. He is described as a generous-natured rugby player who had badly injured his leg and who before the war had been in a bank. An enthusiastic radio amateur, he had joined the RAF Signals Intelligence Service at the outbreak of war. Incidentally, our community of radio amateurs in Britain was to prove an invaluable reserve, both in Signals Intelligence and Signals proper, as well as furnishing the staff our rapidly increasing number of radar stations. Numerous radio amateurs passed through the RAF radio schools at Cranwell and Yatesbury. Some stayed there as instructors, using the RSGB's *Amateur Radio Handbook* as an official textbook.

Peacetime recruit to the RAF Royce Wilkinson, G4HW, became an outstanding fighter pilot. He led a squadron in France in 1940 and was subsequently posted to the Eagle squadron of American volunteers. He escaped capture when downed over enemy territory and was later appointed to command Britain's 'top-scoring' fighter squadron.

Reg Pidsdey, G6PI, a BBC engineer, flew over Berlin in a Lancaster bomber and made famous disc recordings for commentator Wynford Vaughan-Thomas.

Of the first 1000 RSGB members listed as 'on active service' in WWII, 60% were in the RAF (mostly on technical duties), 14% in the Royal Navy and 12% in the Royal Corps of Signals. Naturally, not all of them lived to see the end of hostilities. Pat recalls

Amateur radio callbook of the period, given to Pat by Syd, G2FWZ, in September 1946. It contained just six pages of callsigns. The page shown opposite includes Pat's details.

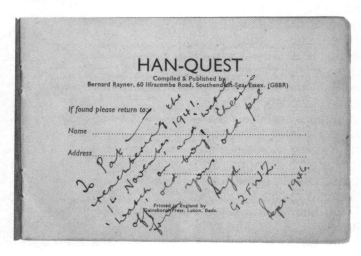

with sadness that about a quarter of the boys in his class at school didn't make V-J Day. The first radio amateurs killed were Jack Hamilton, G5JH, and Ken Abbott, G3JY. They had been drafted as telegraphists onto HMS Courageous at the outbreak of war. They died less than two weeks later, when the ship struck a mine. John Buchan, G4QA, was lost with many other British servicemen in 1940, when the liner Lancastria was sunk while evacuating troops from France. One whose name is remembered annually, when a commemorative award is presented by RSGB, is Norman Keith Adams, G5NM, a young London solicitor who was killed in a flying boat crash whilst gathering electronic intelligence in the Mediterranean. Lew Nash, G4DA, died in Crete when he stayed behind to destroy vital equipment and was ambushed, while Wing Commander John Hunter, G2ZQ, died in the Far East. There were many more.

Inevitably, radio amateurs were amongst those taken prisoner of war. Right under the noses of the guards, a number of secret radio receivers were constructed in PoW camps by radio amateurs. In a BBC 'Open Door' TV programme in 1979 Tom Douglas, G3BA, described some of his work making a receiver whilst a prisoner in the Far East.

Herb Dixon, ZL2BO, a Lieutenant in the RNZNVR was taken prisoner after the fall of Hong Kong in early 1942. During the months that followed, he played a leading role in making and using three receivers in prison camps on the island. To convert an old broadcast receiver smuggled into the camp, flux was made from pine-gun scraped from firewood. This involved careful examination of every piece of wood taken into the camp. Solder was scrounged from the tag boards of old power equipment and prisoners do-

G3SR	Edwards, John, 10, Oak Road, Sale, nr. Manchester.
G3TK	Wood, Ronald, 25, Kenwood Avenue, Leigh, Lancs.
G3VA	Hawker, Pat, Redfields, Whitecross Lane, Minehead, Somerset.
G3WD	Ward, Bert, 90, Other Road, Redditch, Worcs.
G3WG	Wild, John, Springbank, Whitefield, Manchester.
G3XJ	Buick, Ernie, 84, Oatlands Drive, Slough.
G3YX	Pryor, George, The Lodge, Booth Hall, Knutsford, Cheshire.
G3ZC	Cockrem, Hugh, Calstock, Cornwall.
G4AG	Turner, Fred, Greenbank, Westwood Avenue, Brentwood, Essex.
G4AU	Gover, Alan, 30, Amburcote Close, Grove Park, S.E.12.
G4BU	Draper, Dick, 4, Cliff Cottages, Bracebridge Heath, Lincoln.
G4DN	Hamilton, Jack, 48, Queens Road, Preston.
G4DR	Urquhart, Pat, 7, Padwell Lane, Thurnby, Leicester.
G4GM	Ironfield, William, 101, Westwood Street, Accrington, Lancs.
G4GS	Eastwood, Harry, 11, Harper Street, Oldham, Lancs.
G4KG	Spriggs, George, 28, Almorah Road, Hounslow, Middlesex.
G4LO	Fish, George, Lancaster Cotts, North Wold, Thetford.

nated 300 torch cells to provide an HT supply. At another camp, a receiver was built from scratch. An abandoned Austin 7 provided the wire, nuts and bolts, while the rim of the horn provided the vernier tuning dial. Headphones were built from small tin cans, while the graphite from pencils was used to make resistors. Sheet metal and a 4-inch nail were transformed into a variable capacitor and over the course of seven months enough tin foil and thin wrapping paper was collected to make three smoothing capacitors. Perhaps the most remarkable part of the receiver is how the valves were obtained. A small supply was hidden in the operating theatre of a hospital the prisoners were taken when requiring medical attention, but there had to be a good reason to go there. Eventually a prisoner complained of a pain in his stomach and was taken there to have his appendix removed. He returned to the camp minus a perfectly sound appendix, but with three valves concealed in his bandages. On its first trial, in July 1943, the receiver successfully heard the BBC on 9.5MHz, so the news was soon circulating secretly.

Keeping the receiver hidden from the regular searches by guards was not easy. In Shamshuipo camp, the receiver and battery were kept in a water-tight container in one of the lavatory cisterns, permanently submerged in water. At North Point, a hole was dug beneath one of the huts, while a Canadian military band played loudly to drown out the noise.

After a four-hour search on 21 September 1943, one of the receivers was discovered. Along with eight other officers, Dixon then found himself on the receiving end of some harsh interrogation, at the end of which he was sentenced to 15 years imprisonment under appalling conditions. Luckily, he only served two years because of the Japanese surrender in 1945.

Captain Ernest Shackleton, G6SN, who in peacetime was a professional engineer with GEC and a regular contributor of 'Workshop Notes' to the RSGB's *T&R Bulletin* and *The Amateur Radio Handbook* was taken prisoner early in the war and used his 'ham ingenuity' to build a receiver at Oflag 9A/Z, using a valve stolen from a 'talkie' film projector.

WWII left Pat with a strong belief in Churchill's dictum that "jaw jaw is better than war war". He had seen that war brought death, destruction and ruined the lives of people, not only in the UK, but in France, Belgium, Holland and Germany; to people whose prime aim was to live in peace and on good terms with their neighbours. This is reflected in the best traditions of international amateur radio and Pat's belief that there are no winners in war.

Post War

Pat returned to Somerset in September 1946 and resumed his training as an articled clerk at the local chartered accountants but after six years, first in Radio Intelligence first for RSS and then for SIS/MI6 Section VIII, he found it difficult to settle back into what by comparison seemed an unexciting profession. Instead of studying he spent much time on the bands, working DX and keeping in touch with a number of his SCU friends who had similarly returned to civilian life.

Reg Cole, G6RC, relaxing during CW NFD 1948.

When Pat saw an announcement in the *RSGB Bulletin* for an 'Assistant to the General Secretary', he applied, had an interview in London, and was subsequently appointed. He moved to London in September 1947, to lodgings in Chiswick found for him by Reg Cole, G6RC. Reg had been with him at Hanslope Park and Tattenhoe Hostel, and by now had returned to his home in Chiswick. He was a successful DXer and a member of the Barnes and Richmond RSGB Group. Both he and Pat were soon taking part in National Field Day and other local activities.

RSGB Headquarters

When Pat reported to the RSGB offices on the fifth floor of New Ruskin House, Little Russell Street, near the British Museum, he

The Incorporated

RADIO SOCIETY OF GREAT BRITAIN.

THE BRITISH EMPIRE RADIO TRANSMISSION AWARD.

This is to Certify that ...

has satisfied the Council of this Society that he has through Amateur Radio Station ... been in two way communication with Stations situated in twenty-five British Empire Dominion Radio Districts and fifteen British Empire Colonial Areas.

No. 114

Pat qualified for the Britsh Empire Radio Transmission Award in 1947.

found only a small staff working in five, mostly small, offices. There was John Clarricoats, G6CL (General Secretary and Editor), May Gadsden (Assistant Editor), Hazel Lightfoot (clerk), Gwen Thomas (shorthand typist) and Pam and Muriel who looked after the members' address stencils and printed out on an ancient pedal-operated machine that printed the wrappers which the *Bulletin* printers used to despatch the magazine. Later they were joined by a Miss Stanley as book-keeper, but later still lost Pam and Muriel, leaving the creaking wrapper printer to be operated mostly by Hazel. It was all a bit chaotic, as membership rose to some 15,000 at peak. Typical was the small, ramshackle lift that took the staff up to the fifth floor, but also served the lower floors occupied by a major book-publishing firm. It became something of a joke to bring the irascible Sir Stanley Unwin up to the 5th floor when he had entered the lift on the 3rd floor, intending to descend. Pat recalls that he was not best pleased!

Pat acted as a general 'dogsbody', answering members' letters, checking the QSL cards sent for RSGB Awards, accompanying Clarry to Conventions and Official Regional Meetings, reading *Bulletin* proofs and those of the series of small books published by the Society from about 1948 to 1950 and recently reprinted. Gradually he began to write or sub-edit *Bulletin* material, including editorials, and generally to assist in its production – a task that interested him most. He also became interested in the design of display advertisements for RSGB Publications and RSGB events etc. In 1950 he was appointed Assistant Editor.

Pat's starting salary was only £300 per annum, which equates to less than £6 per week. This was less than he had earned as a Special Enlistment, paid from the Secret Vote. Moreover, in the SCUs, accommodation, food, uniform, travel warrants etc had been included. Even in the 1940s, living in London on less than £6 a week was not easy, so Pat certainly had to be penny-wise – even parsimonious; a characteristic he later found to be difficult to shake off. 'Make do and mend' has always applied to his amateur radio operations. In winter, he found that attending Sunday afternoon concerts at the Royal Albert Hall in the upper, promenade area, or the gallery seats in the Lyric Theatre in Hammersmith, home of many excellent plays, cost less than running the gas fire at his lodgings.

Pat spent time learning as much as possible about the craft of technical publishing and writing, typography etc, digested from the weekly pages and supplements of WPN (World's Press News) and books borrowed from or seen at libraries. He acquired copies of *Fowler's Modern Usage*, *Roget's Thesaurus* and *The Oxford Rules for Printers and Authors*. He also bought and studied the HMSO publication *Plain Words*, by Sir Ernest Gowers. Originally intended for the Civil Service, this book contains much good advice and became a best seller as a guide to clear writing. In fact Pat still has a large exercise book full of handwritten hints

Pat, his mother and his sister Betty, on the promenade at Lynmouth in about 1948.

and tips on the preparation of copy, 'casting off', page layout, preparation of scientific papers, cuttings of display type faces, etc. In those days of monotype and linotype printing, he optimistically noted down George Bernard Shaw's "With 26 soldiers of lead, I can conquer the world." He harboured thoughts of writing fiction.

GB1RS

When Pat arrived at RSGB HQ he found that the one large room, used for Council and Committee meetings (and later serving as Pat's office), housed a large 500-watt military-type transmitter that had been donated to the Society by Sir Ernest Fisk of EMI. Council had decided that this should be used as the basis of a clock-controlled automatic 3.5MHz frequency standard service trans-

A Bit of Controversy

G3VA's DXCC certificate, which appears on the wall of his shack. Pat acheived 150 countries, all with home made equipment, most QSOs having been made before he joined the RSGB staff.

mitting a short Morse message and long dash every hour. An antenna had already been erected professionally by a firm specialising in yachting equipment above the roof of New Ruskin House, but various ancillary items including the crystal-oven unit, the antenna system tuning unit, the perforated tape unit etc still had to be completed and installed.

Pat, adjusting the EMI-built 500-watt transmitter.

This was undertaken by 'Dud' Charman, G6CJ, and H A M Clark, G6OT, both Council Members and senior engineers at EMI. Pat had known 'Dud' at Hanslope Park, where he had been responsible for the design of the highly efficient wide-band distribution amplifiers. H A M Clark had, like 'Dud', been an RSS VI Group Leader. But in those days, and for many years later, any mention of RSS was taboo.

Some evenings Pat would stay behind to let them into HQ, take a meal with them at a nearby cafe, and then assist, or at least offer to, in the installation and testing of the equipment.

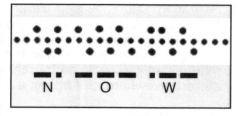

A section of the tape used for the automatic transmission of Morse. Dots, dashes and letters have been added to show how the characters were formed.

The crystal-controlled transmitter was operated by a synchrodyne (high accuracy) clock, timed to put out a Morse transmission once an hour on a frequency of 3500.25kHz. This frequency was chosen because the Post Office insisted the transmissions be inside the bottom end of the 80m band, rather than on the edge of it. Transmissions were automatic, the 'message' being read from a loop of punched paper tape (similar to that used in Telex machines, but with just two 'data' holes in addition to the continuous row of sprocket holes). Transmissions began in 1949/1950, but the clattering relays apparently made a terrible noise when the transmitter came on. This wouldn't have been so bad if GB1RS hadn't been located in a room that was used for meetings, but which also doubled as Pat's office. John Clarricoats became increasingly infuriated with the noise and had the transmitter switched off when committee and Council meetings were taking place.

Pat recalls entering the office one Monday morning and discovering that over the weekend the frequency control unit that kept the transmitter on frequency had ceased to work and that the transmitter had drifted down in frequency, finally settling below the bottom edge of the 80m band. A few Society members had noticed but the Post Office hadn't, so there were no recriminations.

VIEW OF THE CONTROL UNIT AND E.M.I. 500 WATT
TRANSMITTER

GB1RS

HEADQUARTERS STATION
OF THE
INC. RADIO SOCIETY OF GREAT BRITAIN
New Ruskin House, 28/30 Little Russell Street
LONDON, W.C.1. ENGLAND

We acknowledge with thanks your report on the experimental 3,500·25 kc/s. standard frequency transmissions from GB1RS which operates daily at each hour from 0600 to 2400 local time.

RSGB

GB1RS QSL card. A loop of tape can be seen hanging down on the right of the tranmitter.

A Bit of Controversy

Transmissions ceased altogether after Pat left the Society in early 1951, because there was no one else on the staff to keep the transmitter working. It was then disposed of. Questions were asked at a following AGM where it had gone, but this remains a mystery, partly to avoid embarrasing EMI. Apparently it had been installed elsewhere, but not as a frequency standard station.

George Newnes

By late 1950, as Assistant Editor, Pat was taking quite a large part in the production of the *RSGB Bulletin*, a task he enjoyed, but his salary was still too low to contemplate marriage to Gwen Thomas who by then had left the Society for employment as a secretary at a university college in Kensington. He began to look around.

RSGB shorthand typist Gwen Thomas, who became Pat's wife after leaving the Society.

One result was that Reg Cole, G6RC, as Company Secretary at George Newnes Ltd, a major periodical and book publishing company, introduced Pat to Edward Molloy who headed the firm's Technical Books Department. Molloy was looking for a book editor to take over work on the *Radio Engineers' Servicing Manual*, which had been started by an editor who had then resigned. Pat was offered and accepted the job.

Molloy's department was located at Tower Hose, an imposing building near The Strand in Central London. Pat spent several months regretting that he had left RSGB, because initially he found himself in a small office with a perennially miserable middle-aged woman editor. After some months he moved to a larger, more convivial office that before long included a youthful John Reddihough, who became Pat's first assistant and later co-editor.

Pat's work on the *Radio Servicing Manual* went very slowly. Practice in the department for multi-contributor books was to commission articles in, lightly edit them and pass them to the printers with a minimum of delay. Pat found himself confronted with little more than a disparate selection of manufacturers service manuals. Furthermore, he looked at the proofs of the associated *Television Engineers' Servicing Manual* and found that many of the circuit diagrams had no accompanying list of component values – the Advisor Editor (for both books) had advised that since components were colour-coded, such lists were unnecessary.

Pat was convinced this was a serious error but doubted his ability to convince Molloy of this, so he devised a 'cunning plan'. He prepared a carefully worded questionnaire enquiring what information would be most useful for servicing receivers, and sent this to an SCU friend, John(ny) Bowers, G4NY, who, with another SCU operator, Roy Wilkins, G2ALM, had established a radio and television retail business in Worthing. [Later also, as B&W, manufacturers of very high-quality loudspeakers].

Johnny came up trumps. His detailed reply was exactly what was needed to convince Molloy that component values, valve-pin voltages, cord-drive diagrams and where available chassis layouts should be provided in as near a standard form as possible. Pat set about the task but it took longer than he thought it would. Consequently Molloy became increasingly impatient, as it was planned to release both books at the same time. The television book was held up for months and at one stage Molloy suggested to Pat that he might be happier in another job, 'But I survived', muses Pat, adding that soon after the books were eventually published, with the radio book getting good reviews, he got a pay rise. The two books were also sold by the Newnes Subscription Books Division very successfully as a package, under the title *Radio & Television Servicing* (*RTS*) and henceforth this became the policy. Pat was put in charge of future editions of both books and later all the department's books on electronics; suggesting and planning new titles etc.

In December 1952, about the time when *RTS* was finally published, Pat and Gwen were married. Through university connections Gwen had found a pleasant, rent-controlled first-floor flat that was about to become vacant in Gordon Square, Bloomsbury, an elegant part of Central London. It was not far from RSGB HQ with which, as a member of several committees and occasional contributor to *The Bulletin*, Pat had remained in contact.

Pat and Gwen's fist child, Philip, arrived in 1955. He was followed by Virginia, in 1958. Less happily, April 1958 saw the end of the wartime Rent Control Act. Within days, a knock at the door resulted in Pat and Gwen receiving a formal six-month notice to vacate their flat in Gordon Square, without the option of negotiation. The residential buildings on that side of the square were required by the University of London – and have been occupied by them ever since. In October 1958 the family moved to the

A Bit of Controversy

house in Dulwich that is Pat's home to this day.

At the time Pat was busy compiling the 7th edition of *A Guide to Amateur Radio*, a task he had taken over from John Clarricoats and completely revamped. It was to continue in the same basic

Pat's house in suburban South London.

form until it reached its 19th – and final – edition in 1983 (50 years after it was first published). Meanwhile, at George Newnes, together with his young colleague John Reddihough, Pat was busy compiling yet another of the annual volumes in the *Radio & Television Servicing* series, a new edition of the *Radio and Television Engineers' Reference Book* and sundry other titles.

On the bands

You might imagine that having to juggle so many balls at work and having a young family, Pat would have had little or no time to actively pursue his hobby, but his amateur radio log book tells a rather different story. Between January and August 1958 he worked several hundred DX stations on 21MHz CW, including completion of WAZ (Worked All Zones), although most Zones had been worked in 1946-47 from his mother's house in Minehead. His station consisted of a home-made communications receiver based on the pre-war Tobe tuner (which covered only the 1.8, 3.5. 7 and 14MHz amateur bands) plus a 21MHz crystal-controlled converter with its output on 7MHz, both based on American metal octal valves. His home-built transmitter (about 75 watts input) used two of the still popular beam-tetrode 807 valves in the power amplifier, with a separate electron-coupled 6SJ7 VFO on 3.5MHz. All the equipment was squeezed into an in-built cupboard in the living room of their two-room, first-floor flat. Amateur radio was tolerated by Gwen, who had been told "never marry a Ham" – a prophetic warning.

Antennas were a problem at the Bloomsbury flat. There was no access to the roof, but Pat had a 21MHz folded dipole (with dropped ends), made from 300-ohm feeder cable, stretched across the balcony overlooking the Square. By 1958 he also had a similar but sloping folded dipole at the rear of the building that he had managed to suspend from an upper floor window when the upper flat became temporarily empty in about 1956. By switching between the two dipoles he obtained reasonably good results in both westerly and easterly directions, and also south-easterly towards Africa.

Despite the urban location, electrical interference was reasonably low; but early on he was traced by the GPO as causing TVI to a viewer some 150 yards or so away. This led to a hasty rebuild of his transmitter in a surplus metal cabinet, bought from nearby Proops of Tottenham Court Road. Additionally, the problem was overcome by the fitting of a TVI filter built by, Reg Cole, G6RC, Pat's wartime Special Communication Unit colleague and the man who had found Pat his initial accommodation in Chiswick and introduced him to Edward Molloy.

Later, Pat was found to be interfering with the hi-fi system of his neighbour, James Strachey, brother of writer and critic Lytton Strachey (1880-1932). James was an elderly surviving member of the Bloomsbury Group of artists and writers, and his impressive Leak pre-amplifier, main amplifier and large speakers were just a few feet away from Pat's antenna at the rear of the building. The audio was completely wiped out when Pat pressed his Morse key. Pat implemented a novel solution to this problem, by fitting a pilot light in his operating cupboard and a switch in his neighbour's flat. The scheme was that the neighbour would turn the switch on when he was using his hi-fi. Then, if Pat was using his rig, the pilot lamp would light. This proved a success, particularly as the elderly neighbour rarely used his hi-fi system, indeed he told Pat he found it gave him a headache.

Even by 1958 standards, Pat's equipment was dated. Nevertheless, when he re-opened G3VA at Dulwich on 21 November 1958 it remained much the same for many years. This time, however, a 21MHz folded dipole could be suspended between the house and a large Sycamore tree. Later it was joined by a 14MHz folded dipole.

1958 also saw the birth of 'Technical Topics', so it worth reviewing the state of amateur radio in Britain that year.

It was clear that by 1958, the technology of amateur radio was on the cusp of profound change. Already, the large rack-and-panel transmitters were beginning to give way to fully-en-

A Bit of Controversy

closed table-top rigs, primarily to achieve better screening in order to reduce harmonic radiation. It was particularly important to reduce the third harmonic of 14MHz transmissions, because at that time BBC television in the London area was broadcast on

The Italian-made Geloso VFO.

42MHz. TVI drove many amateurs to give up the hobby in the 1950s, but gradually the situation improved. In 1953, Louis Varney, G5RV described the construction of the 'Elizabethan', a near TVI-proof HF 150-watt (DC input) table-top transmitter. CW still remained the main mode of most amateurs for 14MHz DX.

For 'phone, high-level amplitude modulation ruled the roost, requiring a modulator with an output of approximately half that of the RF power-amplifier, i.e. 75 watts of AF to modulate a 150-watt RF power-amplifier. This called for heavy audio transformers, chokes and push-pull Class AB1 designs.

A real heavy-weight, the Labgear LG300.

By 1958, table-top transmitters were available from Panda, with a low-power 'Cub' and a new 'Explorer' on offer. For home-construction, there was the Italian Geloso VFO and the G207 – and later G209 – transmitters, both marketed by KW Electronics, the firm set up by Rowley Shears, G8KW, and Ken Ellis, G5KW. They were also producing their 'Vanguard' 50-watt transmitter in both kit form and as a completely assembled unit (priced 58 Guineas). Geloso equipment was also available from other UK firms.

The year saw the introduction of the relatively high-power LG300 MkII transmitter, a sturdy table-top design in two heavy units: a five-band RF transmitter with a single 813 PA, plus a companion PSU/modulator. Both units were built like battle-ships, by Labgear of Cambridge, a Pye affiliate headed by Sant Kharabanda, G2PU. The transmitter was priced 66 Guineas.

There was by 1958 growing (but still minority) use of SSB, largely with home-built filter or phasing type designs. SSB operators controversially claimed a "9dB advantage" over AM, but recognised it needed more stable receivers, preferably fitted with product (mixer) detectors. Eddystone were offering their Model 888A amateur-bands receiver (six bands from 1.8 to 28MHz) with a mixer detector for SSB based on a double-con-

The Eddystone 640 - made in 'The Bathtub'.

version design with a 1st IF of 1620kHz and 2nd IF of 85kHz plus, for CW, an AF filter peaking at 1kHz. The list price was £110, a significant sum at a time when few earned £1000 a year, and many considerably less.

Eddystone, a brand name established in the 1920s by the Birmingham hairpin and costume-jewellery firm of Stratton & Co Ltd and run by A C Edwards, G6XJ, had, during WWII, after being bombed out of central Birmingham, moved into the 'Bathtub' a former lido in Alvechurch Road. In the post-war period Eddystone produced a series of communications receivers, including the popular 640, although undoubtedly suffering commercial pressure from the many excellent, if heavy, military-surplus receivers that became available to UK amateurs at relatively low prices. In the 1960s Eddystone was to be taken over by the Marconi company, for which it had for some years been making re-badged receivers.

Military-surplus receivers still in common use today included the heavyweight AR88D and AR88LF, designed in 1941 by RCA. They were arguably one of, if not the best, general purpose receivers built in the valve era and were exported primarily to the UK, with few ever appearing on the US surplus market. Others have awarded that accolade, at least for CW, to the National

A Bit of Controversy

One of the best, the RCA AR88D receiver.

HRO, with its excellent crystal filter, originally marketed in the USA as early as 1935. Various versions remained in production up to the professional HRO500 solid-state design in the 1960s. There were also numbers of RCA AR77s, Hammarlund Super-Pros and the military BC342 and BC378 series of receivers, some souped-up by the addition of a Q-multiplier stage or converted to double-conversion by adding the compact BC453 receiver with its 85kHz IF as a 'Q5-er'.

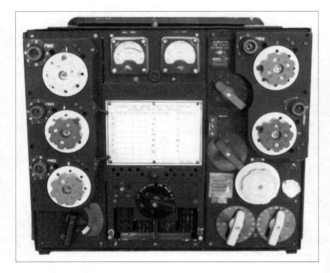

Thousands of these T1154 transmitters were used in WWII.

British-made surplus models included the R1155 series, with the associated T1154 transmitter. These had been built by the thousand from June 1940 by Marconi and were used in all the aircraft of Bomber Command, as well as many fighter-bombers, flying boats, reconnaissance aircraft, ground stations, air-sea rescue launches, etc. They were based on the 1937 civil aircraft design AD67/AD77, which were used pre-war in the Empire flying boats. One of the best British communication receivers was the Marconi CR100 (B28), which came close to the performance of the National HRO and RCA AR88.

British, Canadian and American companies produced several versions of the 19 Set, although this design had been subjected to last minute changes, included poor quality components and had proved far from reliable when used in North Africa.

The ubiquitous 19 Set.

A revolutionary British receiver, the Racal RA17, was first produced about 1957-88, but at a cost that limited its use by amateurs. The design was based on the Wadley-triple-mix configuration, making it virtually drift-free. It also featured a film-tuning dial that could be set to within 1kHz on all of its thirty 1MHz-wide bands. Although not designed for SSB, it provided excellent performance

With its film tuning dial, the RA17 could be tuned accurately to within 1kHz across each of its thirty 1MHz bands.

on this mode when carefully adjusted (an add-on SSB unit was later manufactured, but seldom fitted).

The WWII surplus market had passed its peak by 1958, although plenty of material could still be picked up at low-cost. For London amateurs, Lisle Street (of dubious reputation!) still remained a Mecca. From Pat's flat in Gordon Square, Proops in Tottenham Court Road was a short walk, although purchases were limited by the restricted space. In 1958, on offer from Proops was the W/S No 36, a complete 50-watt 10-60MHz (three bands with plug-in coils) 'table-top' AM transmitter. With its power supply it was offered at £12/10/0 with the claim "Get this new rig on the air in five minutes flat." Henry's in the Harrow Road was noted for its low-cost stocks of surplus crystals, while P C Radio in the Goldhawk Road was another source.

By far the most significant development of the era was the marketing in the USA in January 1957 of the first SSB/CW transceiver, the Collins KWM-1, intended for fixed or mobile operation. It was this model that

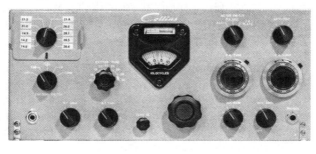

With its SSB capability, the Collins KWM1 ushered in a new era in amateur radio.

set the pattern of amateur equipment for decades to come. Although the KWM-1 with its Collins mechanical filter was not widely used in the UK, on account of its price and rather limited frequency coverage, other manufacturers were soon offering HF SSB/CW transceivers that were basically patterned on the pioneering Collins designs. These included, in 1958, KW Electronics with its design patterned on the KWM-1 and KWM-2. Daystrom also launched a series of Heathkit SSB/CW transceivers in both kit and assembled form.

VHF operation on 70MHz, 144MHz, UHF and above still depended almost entirely on home-built equipment or modified 'surplus' military equipment, such as the American SCR522. People primarily used AM, but there was also some CW. There were several amateur TV (ATV) enthusiasts. Mostly, VHF transmitters were crystal-controlled for a single-channel. Many VHF/UHF amateurs made use of low-noise converters in front of HF receivers, early on often using the surplus RK25 series of airborne converters. Pat is not sure whether the RCA low-noise Nuvistor miniature valves such as the 6CW4 were available in 1958, but when available they soon became popular for the pre-amplifier stages.

Wilf Allen, MBE, G2UJ, did much to encourage VHF operation in his early post-war period contributions to the *RSGB Bulletin*, although by 1958 his monthly column had been taken over by Fred Lambeth, G2AIW. Wilf had been one of the CWR/RAF 'Early Birds' going to France in a wireless intelligence unit on the outbreak of WWII and later doing covert wireless work in Sweden. W A (Bill) Scarr, G2WS, described a number of home-constructional VHF projects. Neither was a professional engineer: G2UJ was in insurance and G2WS was an education officer and former teacher who later went to the Sub-Continent on behalf of the British Council, before retiring to Weston-super-Mare. George Jessop, G6JP, with a career in the valve manufacturing industry, became a prolific source of HF/VHF/UHF designs.

Propagation research was given a boost in 1957-58 by the International Geophysical Year (IGY), with Dr Smith Rose securing a role for radio amateurs, UK teams were co-ordinated by Geoff Stone, G3FZL.

On 4 October 1957 the USSR had launched Sputnik 1, with Sputnik 2 following on 3 November. Their 20 and 40MHz signals had been heard by many amateurs. Sometimes signals were received at unexpected times, seemingly verifying transmission in the 'whispering gallery' mode. Charlie Newton, G2FKZ was one of the prominent IGY activists.

Although by 1958 mass production consumer electronics was largely based on printed circuit boards, with transistor radios reaching the US market in late 1954 and the UK market from 1956, amateur radio equipment was still firmly based on hard-wired construction and all-valve technology. An exception was the emerging use of silicon diodes for power rectification, although for a time a string of diodes was needed for high-voltage supplies. Mercury-vapour rectifiers were still used in the highest voltage supplies, while the 5Z3 rectifier was popular for medium-voltage.

The first germanium point-contact transistor had been demonstrated in the Bell Telephone Laboratories in New York in December 1947 and publicly announced the following July, but at first received only a muted response. Even with the later development of the alloy-junction PNP and NPN devices, industry responded only slowly. The early devices proved difficult to manufacture. With a wide variation in the characteristics even between similar types, they mostly provided gain only at AF and MF and at very low output power. For a few venturesome amateurs they of-

The rectifier valve of choice for many in 1958 was still the 5Z3.

fered a field for experimentation rather than serious operational use, but progressively the semiconductor industry began to emerge, with transistors seen as a promising rival to thermionic valves.

In 1958, amateur stations often comprised many auxiliary units and test instruments, usually home-built. Typically, these included PSUs, VFOs, ATUs, SWR meters, crystal calibrators, crystal activity meters, valve-voltmeters, grid dip oscillators (for which, even today, valves tend to be still superior to FETs or bipolar transistors), electronic keyers and semi-automatic 'bug' keys, and absorption wavemeters. Better-equipped stations included factory-built or occasionally home-built oscilloscopes and signal generators. How different from today when transceivers are virtually self-contained stations, often with in-built 'auxiliary' facilities including automatic ATUs of limited range.

Just as today, in 1958 the key to successful DX operation, with relatively low power was the antenna. Virtually all the main fundamental types of antenna used by amateurs had already been developed but choice was, as now, often limited by considerations of the available space, cost, appearance, safety and local planning restrictions. Most simple wire antennas such as dipoles etc were home-constructed and cut to resonance. Yagi and Quad antenna elements were available on the market, but 'plumber's delight' Yagi antennas were often home-built.

The kind of leaflet that might drop out of any electronics magazine in the 1950s or 60s.

Meanwhile, back at work

The *Radio & Television Servicing* book grew so popular that it sold-out very quickly. It was revised, split into four volumes and re-released. Every year a new volume was added, until there were seven. At this point it was becoming too bulky and too expensive, so old volumes were compressed and combined to maintain a six or seven volume set. Working to an extremely tight time schedule, it was a continual hectic process to get the new edition out before the end of the year so that it could be considered 'next year's' edition.

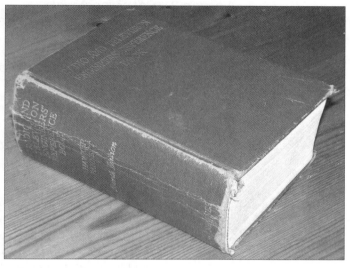

Pat's favourite work was the *Radio & Television Engineers' Reference Book*, which is a weighty tome indeed. Almost 3in thick, it weighs close to 4lbs. Originally it came out under the name of Molloy and Pannett. A large number of people contributed to this book, some of them radio amateurs. With over 50 specialist contributors and 47 Sections, to describe it as 'encyclopaedic' would be somewhat of an understatement. It was planned and compiled by Pat, with some help from W E Pannett, a retired Marconi engineer. The following is the contents list of the 4th edition.

The Radio and Television Engineers' Reference Book, published by George Newnes.

Section 1. *Formulae and calculations*
Section 2. *Communication Theory, Electron Optics and Colour Television*
Section 3. *Materials*
Section 4. *Studios and Studio Equipment*
Section 5. *Transmitter Power Plant*
Section 6. *Broadcasting Transmitters*
Section 7. *Communication Transmitters*
Section 8. *VHF Transmitter-receiver Equipment*
Section 9. *Amateur Radio Equipment*
Section 10. *Television Transmitters*
Section 11. *Transmitting Aerials*
Section 12. *Radio Frequency Transmission Lines*
Section 13. *Waveguides*
Section 14. *Broadcasting Receivers*

A Bit of Controversy

I am going to select three sections of this book for special attention, because they are undoubtedly sections that Pat would have had a lot to do with the production of.

Section 9. Amateur Radio Equipment (written by Pat)
Licences – Amateur Frequencies – High-frequency Transmitter Design – Oscillators – Exciters – Power Amplifiers – Television Interference – Modulation and Keying - High-frequency Receiver Design

At 20 pages in length, this is one of the shorter sections of the book. It begins with a description of what amateur radio is about.

This is followed by a table of the amateur bands, power levels, permitted modes, etc. You can see that it is basically the same as the frequency allocation table of today, except that we now have more bands but a smaller allocation at 430MHz.

A table of the characteristics of numerous RF tetrode Power Amplifier valves of the day includes several familiar models, such as the 807, 813 and 6146.

The generation of single sideband is discussed, as is frequency modulation (which was rarely used on the amateur bands in 1963).

The list of requirements of an HF receiver for amateur use is practically the same as it is today; good sensitivity and signal-to-noise ratio, a high order of selectivity, freedom from spurious responses, electrical and mechanical stability, freedom from mains hum, accurate calibration and the convenient positioning of controls.

A number of the illustrations in this section look as though they came straight from RSGB publications, indeed the bibliography at the end of the section mentions *A Guide to Amateur Radio* (edited by Pat) and *The Amateur Radio Handbook* (edited by John Rouse).

Section 22. Communication Receivers (written by a Marconi engineer)
Requirements – Sensitivity – Spurious Signals – Signal-to-Noise

If you repaired televisions in the 1960s, you are guaranteed to have seen this book and probably also this flier.

Ratio – Oscillator Tracking – Intermediate Frequency Amplification – Automatic Gain Control – Typical Receiver

This is another relatively short section, yet the author manages to describe just about everything that was important about a receiver at the time. Moreover, although almost half a century has elapsed since it was written, 'The Requirements for a Communications Receiver' have changed very little. As the book says; "Operation should be simple", "several degrees of selectivity should be provided", "spurious signals of all kinds should be negligible" and "the component parts should be reliable under extreme conditions of temperature and humidity". Moreover, techniques being exploited by only a few radio amateurs today were mentioned then; such as diversity reception. One or two 'requirements' have inevitably fallen by the wayside though; "access should be good for maintenance purposes", and "the receiver should be comparatively cheap to produce" are seen as irrelevant, whilst the use of frequency synthesisers has eliminated drift.

Calculations for signal-to-noise ratio are included in this section, as are the characteristics of coupled circuits. There is even a design for a double-conversion communications receiver, complete with all the component values. It would have made a decent short wave communications receiver, but bear in mind this wasn't an amateur radio book.

Section 47. Progress and Developments (compiled by Pat)
Tape Recorder Developments – Thermoplastic Recording – Scan Magnification – Transmitter Developments – Stereo Broadcasting – Aeronautical Radionavigation – Mechanical and Piezo-electric Filters – Triple Conversion Tuning System – Planar and Epitaxial Transistors

The edition of the *Radio & Television Engineers' Reference Book* that I had access to whilst researching this book was the fourth, printed in 1963. I think it is no coincidence that in the same year KW Electronics started to manufacture the KW2000 series of transceivers, which used a mechanical filter. The mechanical filter depicted in the section of the book is a Collins, although KW used a Japanese equivalent made by Kokusai. Although mechanical filters are now a little smaller, Collins still manufacture them, indeed they find their way into a number of items of commercially-made amateur radio equipment.

Although no callsigns (apart from Pat's and that of P Jones, G2JT, of Aerialite Ltd) are given in the list of specialist contributors in the *Radio & Television Engineers' Reference Book*, one of the

Another one of Newnes' fliers.

names will undoubtedly be familiar to many radio amateurs – that of Les Moxon, G6XN. From the book, here is Les' resume:

"L. A. Moxon, B.Sc(Eng.), A.M.I.E.E., was educated at Clifton College and the City and Guilds Engineering College, obtaining the London University B.Sc. degree in 1929. After two years research under the auspices of the D.S.I.R. he joined the staff of Murphy Radio Ltd., where he was responsible for development and research in connection with broadcast reception. In 1941 he joined H.M. Signal School, Portsmouth, where he was concerned with the development of radar receivers. He is now a member of the Royal Naval Scientific Service." Les would go on to write several books for RSGB, to develop some interesting antennas and corresponded with Pat on antenna and propagation topics frequently for 'TT'.

Between some sections there were advertisements from companies long gone and forgotten; companies such as Lustraphone, Wandleside Cables and Vectron. Other more familiar advertisers included EMI, Marconi, Ernest Turner, S G Brown and AVO.

The *Radio & Television Engineers' Reference Book* was advertised using fliers that were inserted into the kind of magazines that interested parties would be likely to see already, including the *RSGB Bulletin*. Newnes also used door-to-door salesmen to sell their subscription books, which led to trouble because some salesmen made promises they shouldn't have. Newnes eventually became part of the Odhams group, who also published the Sunday newspaper *The People*. Ironically, *The People* were running a campaign about the unjustified claims being made by some door-to-door salesmen. This led to Newnes hurriedly disposing of their subscription books division. The titles were sold to Pergamon Press, which had been founded by Robert Maxwell. Maxwell wanted to continue producing the titles and his firm asked Pat to help, but his request was declined. Consequently, the 1965 edition was the last one he had anything to do with. Pergamon didn't produce any further editions of *RTERB*, but Amos of the BBC later produced a completely different edition for Butterworths. The book is no longer in publication, but if you have an interest in early radio or television and see a copy at a radio club sale or a rally, you might like to consider buying it.

It seems to be the nature of the publishing business that companies merge, titles are sold on and you keep bumping into the same people at different places. As Pat says, "Molloy always had a motto: Get the first edition out! If there are mistakes we can always correct them in the second edition." And there were some indeed in the 'Formulae and Calculations' section. The professor who pointed them out was commissioned to go through the section in detail, on the basis that he would be paid £1 for every mistake he could find. Apparently he ended up earning himself about £40. On the whole they were minor mistakes.

Moving on

In *Radio and Television Servicing* they were very often using circuit diagrams direct from the manufacturers, but even these

Members of the radio press at an amateur radio exhibition in the 1960s. Left to right: Pat Hawker, G3VA (*Electronics Weekly*), John Rouse, G2AHL (*RSGB Bulletin*), Austin Forsyth, G6FO (*Short Wave Magazine*), John Wilson, G3BGP (*Electronics Weekly*), John Clarricoats, G6CL (*RSGB Bulletin*), F L Devereaux, ex-5FA (*Wireless World*).

weren't perfect. On at least one occasion the publishing house was threatened with legal action, because a circuit error was reckoned to have resulted in a set of valves being damaged. "It was a very hectic time", says Pat, "but very much an office job and I very much preferred *Electronics Weekly*, getting out and about and overseas visits."

Electronics Weekly started in 1960 as a weekly newspaper type publication. The first editor was Cyril Gee, who had come from the Mullard Press Office. It was set up by National Trade Press, an affiliate of George Newnes Ltd. Not long after its inception, Pat started writing for it on a freelance basis. By 1963 the editorship had passed to John Wilson, G3BGP, and it was he who head-hunted Pat, knowing of course that he was already writing 'Technical Topics' for RSGB. In fact it was more of a transfer than anything else, because it was still part of George Newnes Ltd.

Consequently, Pat transferred from the Technical Books department at the beginning of 1963. In *Electronics Weekly* (*EW*) there were plenty of pages to fill every week, but being the only periodical of its kind at the time they received all the press releases and lots of invitations to launches and visits. Pat was to stay there until October 1968, although he still freelanced for a couple of years on *Radio and Television Servicing* with John Reddihough in charge.

EW covered all aspects of the industry; new developments, company and commercial news, Government and Party politics; in fact it aimed at competing with the *Financial Times* - but only in

A Bit of Controversy

the field of electronics, including communications, broadcasting, radar, computers and aeronautics.

In the amateur radio world, the advances that Pat had foreseen in the 1950s were now taking place. Introduced in 1963, the KW2000 was fitted with a Japanese mechanical filter, marking not only a major development but also, for the first time, heralding the coming influence of Japanese products on the amateur radio market worldwide.

Introduced in 1963, the British-built KW2000 HF transceiver.

It would not be long before Japanese SSB/CW transceivers were being marketed in Europe under the brand names of Trio (later Kenwood), Sommerkamp (Yaesu) and Icom. Things would never be quite the same again, with long-established American firms gradually pulling out of the market or transferring production to the Far East.

Getting about

In 1963, 1965 and 1967 Pat went to Switzerland, to attend the technical side of the Montreux Television Festival. He visited Paris at the

The colour TV conference in Berlin, 1965.

personal invitation of Henri Peyroles of CFT to see work on the SECAM (SEquential Colour And Modulation) television system, which was being proposed by the French. Pat also went to a conference in Berlin, associated with the PAL (Phased Alternate Lines) colour system developed by Walter Bruch of Telefunken.

In 1965 in Los Angeles, the Hughes Aircraft Corporation was visited as part of the Early Bird communication satellite launch programme. Pat describes this as "A fantastic trip". At the time there was a terrific controversy going on, with the

Post Office and various other British organisations wanting to set up a global system of three communication satellites in 12-hour orbits. Being in lower orbits, the propagation delay would have been much less than the delay sending signals via satellites in geostationary orbits, plus the signals would have been stronger because the satellites would have been physically closer to the ground.

The Early Bird satellite undergoing testing in California.

Dr Harold Rosen, one of the three Hughes Corporation scientists who developed geostationary satellites, was dispatched to Britain to try and convince the Post Office and international authorities that geostationary satellites would be a more satisfactory system, because with satellites in 12-hour orbits it would be necessary to have takeover periods and multiple, tracking dishes. For a long time the Post Office argued that the propagation delay – which is over 120milliseconds – would prevent people from being able to

SEMICONDUCTOR FUTURE

THE wholehearted acceptance of semiconductor techniques has been one of the contributing causes to the vast strides made in the Japanese electronics industry during the past decade, so that the views of Dr Yukimatsu Takeda, of Nippon Electric Co, on current trends in semiconductor technology are not without interest.

First, he suggests, silicon will continue to make inroads on germanium. In the very near future, he believes, transistors and diodes will be mostly of silicon. Epitaxial technique "will be the most important crystal fabrication process with planar technique predominant in device fabrication". With a combination of these techniques, semiconductor devices capable of "several hundred watts in output power" and "frequency characteristics to several gigacycles" will be "not too remote".

The successful application of thin-film evaporation techniques is greatly spurring development of solid-state integrated circuitry. While reliability and microminiature dimensions have already been achieved, "it will not be easy to accomplish low-cost production", says the doctor. The yields of present day solid-state integrated circuitry, in his view, is woefully low. Once successful high-yield production has been achieved,

emphasis will shift to the design of chip patterns and the development of good fabrication techniques.

Progress is surface passivation techniques, he considers, is effecting changes in the encapsulation of devices. He points out that several non-Japanese firms are using epoxy resin moulding and his own company has developed ceramic-glass sealed encapsulation for low cost packages. This hermetic ceramic packaging lends itself to mechanisation in device manufacturing and also to automation in circuit assembly, he claims.

Progress in thin-film techniques is also resulting in other versatile developments such as field-effect and metal base transistors. The low noise and high output impedance of field effect transistors point to their use before long "in many practical applications". The metal base transistor, he feels, will "require several more years of research work to complete the industrialisation".

Finally, Dr Takeda points out that semiconductors are coming to the fore in "optoelectronics", for example in the injection laser. Research and development in this field of application, it is suggested, will remain for years as one of the "most wanted topics", encompassing important and diversified advancements of semiconductor technology.

NEW THINKING

12/2/64

Pat wrote the vast majority of 'New Thinking' in *EW*.

Colonel John H. Glenn

Luncheon

*

Presiding

DR. MATTHEW W. MILLER,
Chairman, 3M Research Ltd.

*

ABRAHAM LINCOLN ROOM

Tuesday, 31st May, 1966

While working on *Electronics Weekly* Pat attended a lunch at the Savoy Hotel at which Colonel John Glenn, the first American to fly in space, was present. Pat's menu card is signed by John Glenn himself.

John Glenn.

hold conversations, because hearing their own voices coming back, delayed, would confuse them. It was an argument that went on for months and months, but Dr Rosen managed to convince Pat that the real problem with the propagation delay was that the echo suppressors the Post Office were using weren't good enough. He said that if the Post Office improved the echo suppressors they would find that the apparent effect of one's own voice coming back would disappear. In March 1965 Hughes organised a press visit to their space labs in Culver City "to bend people's ears and jolly everybody up". This was the trip that Pat went on, and he was the only technical journalist on it. He was later told that it cost the Hughes Corporation about $20,000 per person to organise the visit, but as Hughes were partly the owners of the airline that had taken everyone there, perhaps they got special rates on the seats! Among the 'jollies' was a day at the original Disneyland at Anaheim and the West Coast Premiere of 'The Sound of Music'. More serious were visits to the Hughes Research Laboratories in Malibu on the Pacific Coast, birthplace of solid-state lasers.

On the way home, Pat stopped for a day in New York to look over the colour TV studios of RCA. At that time there was another major controversy, surrounding which colour television system ought to be adopted on this side of the Atlantic. Pat had arrived at *Electronics Weekly* just in time just in time for this to break into the press, but when he got there almost the first thing he read was a convincing piece in the Decca newsletter on SECAM. Not really knowing about the battle that was going on, Pat wrote an enthusiastic feature on SECAM, pointing out its advantages (which, incidentally, Pat says it still has). It was around this time that the facetious acronyms appeared: Never Twice the Same Colour for NTSC, the American system, and Systems Europeans Contra les AMericans for SECAM. Later, Walter Bruch came along with the PAL system, which wasn't even in the original European Broadcasting Union's scheme of things. It became known as Pray And Learn.

Apparently a system similar to PAL had been proposed in America by the Hazeltine Corporation, but it was discounted because every receiver would need a delay line and at that time the only delay lines capable of satisfying the needs of the PAL system would have been made from glass. For PAL, Telfunken in Germany worked like fury and managed to make a delay line made from steel, which was much cheaper than a glass unit. When Pat went to Germany they offered to supply any television manufacturer with steel delay lines for well under £1, which undercut the £10 or so of the glass version enormously.

A Bit of Controversy

After this Pat returned and wrote a centre page spread in *Electronics Weekly*, saying that "PAL could be the compromise for Europe". This upset the proponents of SECAM, because he had previously been writing in support of them! One of the biggest advantages of SECAM was that video recording it was much more straightforward than NTSC or PAL. The standard 2-inch video recorder used for recording monochrome television was massive and cost about £20,000, but it worked fine for SECAM. In those early days the BBC were very strong advocates of 625-line NTSC and worked with Ampex to develop a colour video recorder that used their 2-inch tape and helically scanning heads. Incidentally, this difficulty in recording NTSC is why colour television had been slow to take off in America, because the country's five time zones mean that you needed to record material in order to play it at the same clock time in the local zones. ABC and ITV had advocated 405-line colour, experimenting with 405-line SECAM, 405-line NTSC and 625-line PAL. It was partly resolved in a conference in Oslo, where Howard Steele, now Chief Engineer at the IBA, represented the ITV. He fell out with the Post Office because in the end around eight or nine systems were being proposed, there being numerous slight variations of the three basic systems.

In France the government appointed a Minister to make a decision. Unsurprisingly he decided in favour of the French system, but it was not disclosed until much later that SECAM is far more difficult to handle in the television studio than NTSC or PAL. Pat believes that to this day the French TV studios run in PAL, but the signals are then converted to SECAM for broadcast. Pat freely admits that all these controversies provided great scope for features. Not that there was a shortage of interesting things to write about, with transistors finding their way into everything and the first integrated circuits being produced. Pat was the Communications Editor, but there was also a Components Editor, an Aeronautics Editor and a Computer Editor (Pearce Wright, who later became Science Correspondent of *The Times*). "We had quite a strong staff", he says.

An early SECAM television manufacturing facility in France.

These days it has been largely forgotten how fierce the TV

standard issue raged, with the BBC, BREMA (British Radio Equipment Manufacturers' Association) and the GPO clinging on to their belief in NTSC.

Electronics Weekly still exists today, but instead of being on sale in newsagents it is now a controlled circulation paper, being sent free to those in the industry. Pat isn't sure if the current members of staff attend lectures at the IEE to learn of new techniques, as he did. One of the mini features he wrote weekly was 'New Thinking', which brought the attention of readers to the forthcoming technologies of the day. Initially this was written by various members of staff, but it ended up with Pat writing 99 out of 100 of them. He still has a folder full of the clippings, a couple of which are reproduced in this book.

Shortly before Pat left *Electronics Weekly* 'New Thinking' was replaced by a technical page, which Pat also wrote. He described his time at the magazine as "It was a very prolific and very interesting time in electronics because of course the British electronics industry was at a fair size. It hadn't been entirely swamped by imports", going on to recall some of the names of yesteryear that have now disappeared; Marconi, Plessey, Racal and AEI. Telecommunications by radio and satellite had been Pat's domain at *Electronics Weekly*, including radio and television broadcasting and satellite comms. Also the first use of digital PCM (Pulse Code Modulation) in telecommunications.

A 'cordless' era — one day

THE semiconductor has brought with it new frontiers in low-power operation with all its resulting changes on the power rating of components and on the feasibility of battery operated equipment of increasing complexity.

One of the immediately important areas in which there is a continuing search for even lower powers, particularly in the quiescent mode, is in the field of low-power personal radios for two-way communication or for paging or alerting.

Several units already make considerable use of various pulsing techniques in which the equipment is on for only a small percentage of the total time. Meanwhile a new onslaught on this problem is reported from across the Atlantic.

Dr James Meindl, of the US Army Electronics Command at Fort Monmouth, recently described to the IEEE new circuit techniques which can result in a five to ten times reduction in the standby battery power drain of helmet radio receivers used primarily by field troops for portable, lightweight radio communication.

Dr Meindl believes that although the performance capabilities of low-power digital circuits have been defined in considerable detail, relatively little information has been reported on the minimum quiescent power requirements of linear circuits.

He suggests that by using carefully formulated optimum design techniques, the quiescent power requirements of low frequency, wideband, tuned and low-noise amplifiers as well as oscillators, mixers and detectors frequently may be reduced more than an order of magnitude.

Using such techniques, it has been possible to extend the lifetime of batteries in a 51 Mc/s helmet radio receiver from a maximum of only about 20 hours right up to 150 or 175 hours.

Dr Meindl anticipates that these design concepts could well prove applicable in such fields as radar and combat surveillance equipment.

The other approach is that of improving the basic power sources. For though the nickel-cadmium chargeable cell, the mercury battery the alkaline-manganese and conventional and heavy-duty Leclanché cells now present a formidable range, there still exists a requirement for improved forms of portable, lowcost power sources, particularly for applications still requiring peak currents rather greater than those in the personal radio field.

Some American writers have suggested that we shall eventually find ourselves truly into the "cord-less" (mains free) era for almost all domestis electric appliances, but this era is one which still needs the dual-pronged approach of using and producing power more efficiently.

NEW THINKING

25/5/6?

Another example of how Pat envisaged the future.

In early 1968, as the sole technical journalist in a press party, Pat was invited to visit Ascension Island, to see a new Marconi-built satellite ground station for Cable and Wireless. Incidentally, he spoke over the 4GHz link to an American amateur at NASA.

Pat believes that the five plus years he spent with *Electronics Weekly* were perhaps the most prolific and in some ways the most satisfying period of his writing career. With plenty of pages to fill and plenty of scope for interesting assignments, he kept himself busy with features, news reports, his 'New Thinking' pieces, contributions to the 'Janus' column, etc. He had been closely involved with the choice of which colour television system would be adopted in Britain and which type of global satellite system would be used.

On the BBC

The BBC World Radio Club was a regular program, broadcast on the World Service, and Pat used to go along to Bush House whilst working at *Electronics Weekly* and record items for it. The programme had developed from the BBC Short Wave Listeners' Club, but had been renamed. In the 1960s the presenter left World Radio Club, which left the BBC looking for a suitable replacement. They put a note in the *RSGB Bulletin*, inviting any amateur who might like to be a presenter of BBC World Radio Club to apply. Pat duly did, but by the time his letter was received they had already appointed someone. The producer at the time was Joy Boatman. She is someone who Pat remembers as "a very talented young lady". Joy sent him a letter suggesting that he did a few short pieces about satellites instead. Pat duly wrote some scripts and went along to record them. Joy then put it to Pat that she would like to try him in an interview type role, rather than reading a script, because, as she said, "When you're reading a script it sounds like you're reading a script". With a chuckle, Pat describes her as "dead right!" The result was that Pat would go along and record a few technical interviews at a time, which would be inserted into programmes and broadcast later.

Henry Hatch, G2CBB, who was a BBC man, used to be prominent on the BBC World Radio Club. Sometimes he, but more often the presenter, would interview Pat. Even after joining the engineering department of the IBA, Pat continued to occasionally record pieces at Bush House.

The IBA

By 1968, although happy working on *EW,* Pat (now aged 46) began to doubt how long he could continue the nomadic life that

Apollo Satellite Communications Station built by Marconi for Cable and Wireless Limited.

Top: Ascension Island is a lonely volcanic rock in the South Atlantic.

Middle: First Day Cover showing the special stamps that were issued on Ascension to mark the opening of the Cable & Wireless satellite terminal.

Bottom: The BBC Atlantic Relay Station, which broadcasts the World Service to Africa and South America, visited by Pat.

took him away so often from his wife and children. His marriage came under strain, eventually leading to a separation that lasted from 1973 until 2004, although all the family remained in close touch and spent time together.

Pat joined the Engineering Information Service of the Independent Television Authority in October 1968, at a time of intense activity for the Authority's engineers. In 1967 the Government had finally decided that ITV would be authorised to establish a colour service using 625-line PAL, but only on UHF channels (Bands IV and V) from a new transmitter network based on sites shared with the BBC and capable of later providing a fourth channel that remained unallocated. The colour network was to open simultaneously with the BBC1 625-line colour service in November 1969, from four co-sited facilities capable of reaching some 60% of the UK population. Meanwhile, the BBC1 and ITV programmes would have to continue to be transmitted on 405-lines in black and white.

Pat's IBA pass card.

This presented a major challenge to the IBA, who apart from regulating the ITV companies was also responsible for the transmitters. The Authority quickly appointed the energetic and ambitious young Howard Steele who, as chief engineer of ABC Studios at Teddington, had been closely involved in the testing and choice of the colour system, as its new Chief Engineer (later Director of Engineering). He expanded and strengthened the engineering staff at Brompton Road, Knightsbridge, but planned to operate the new transmitter network with the same number of field engineers by developing the use of remotely-controlled unattended transmitters, a technique that had already been pioneered on a small scale by ITA for the VHF network. The ITV companies would be responsible for converting their studios to 625-line colour with IBA providing line-standard electronic converters at the transmitter sites to permit duplication of the programmes in 405-lines black-and-white. Higher antenna masts would be required at some sites. Inevitably, with half of the UHF transmitters located at BBC sites, ITV regional boundaries would be changed.

All this would involve the viewers as well as the companies in a major learning process. Howard Steele foresaw that the IBA would need to set up an Engineering Information Service, comparable to the BBC's long-established Engineering Information Department, but he also had other aims; gaining more recognition by the industry and public of the prowess of his engineers and their endorsement of new technology. As a council member of the Royal Television Society, composed mainly of engineers, he had undertaken to find a new editor for their journal that had fallen behind in issues.

He brought to Knightsbridge Alan James, engineer-in-charge of the Caldbeck transmitting station and appointed him Head of EIS. They put an advert in *EW* for two Senior Engineering Officers. Pat noted the advert with interest, but did nothing about it until he received a phone call from Howard Steele asking him if he had seen the advert and might he be interested. If so he would be invited along to Brompton Road one lunch-time. Pat accepted the invitation and after a brief discussion, Barney Keelan – IBA's Head of Information and a Personnel Officer – was brought in as an informal, formal Interview.

Steele raised the question of the RTS journal as a spare time task, but suggesting there would be no objection to Pat doing some of this work in the office. Pat similarly raised the question of 'Technical Topics' and other freelance work. Steele had been secretary of the Imperial College Amateur Radio Society and readily confirmed that Pat could continue writing on amateur radio and non-competitive articles. Pat became the third member of EIS, with a wide range of duties aimed at publicising IBA engineering, dealing with enquiries from viewers and the trade, speaking at a series of meetings with dealers held jointly with the BBC under the aegis of TV aerial firms, etc. One of his first jobs was to write a booklet on colour television, for distribution at a meeting of ITV companies and their advertisers.

At first the 625-colour project seemed to be going well, but problems soon arose. The novel Varian high-power 5-cavity klystrons, initially developed for radar, would not work effectively for TV signals. Tom Robson, then Head of Transmitters, was despatched to California to help sort out the problem. Then in March 1969, the 1,265ft-high cylindrical steel mast at Emley Moor collapsed, depriving the Yorkshire Region of its ITV programmes. A temporary mast from Sweden was quickly erected, but Pat recalls facing an angry audience of dealers at Sheffield, a notably hilly area, complaining that many of their customers could no longer receive the ITV service. A Sheffield relay was soon added, while work on a new concrete mast at Emley Moor was rushed ahead to meet the critical November deadline.

In the end, all four UHF stations came on stream on the due date and the whole changeover project was deemed a great success. Work continued over the next few years to bring the 625-

line service to the other regions but there also remained, as fore-seen, the need for many local relay stations to fill-in the gaps resulting from the change from VHF to UHF. It had been esti-mated jointly with the BBC that this would need some 800 sta-tions. In practice, it required twenty years and 1,000 stations to complete the coverage and permit the closing down of the 405-line network. All this involved Pat and an expanded Engineering Information Service, keeping the trade and public aware of where and when they would get a good colour picture. All relays as well as many main transmitters were operated unattended, without increasing the need for more field engineers than had been used for the VHF service. Eventually the whole system was run from just four regional control centres.

In the early days problems threatened the expansion of UHF coverage, in the building of the more powerful local relays. A con-tract with Plessey for the supply of transposers based on the use of four paralleled 200-watt units with TWT (travelling wave tube) amplifiers went sour when the company found that this configura-tion could not meet the required specifications. A hurried search for an alternative system resulted in the purchase from France of a more conventional transposer design that used a UHF tetrode, but the completion dates of these key local relays was delayed. Dealers in the affected areas were soon complaining.

By now several Engineering Information Officers had been recruited, some from the IBA transmitter staff, some from the ITV companies and some from outside including John Brodkzy, G3HQX, with long experience in the television retail trade. In 1971 John came to Pat with a proposal that IBA should begin a weekly series of on-air announcements to provide information to the television trade, akin to that regularly transmitted by the BBC in sound only. The IBA had never pro-duced 'programmes' other than the trade music broad-cast from the transmitter control room to accom-pany the test signal, so Pat was doubtful if John's pro-posal of announcements over the ITV network could be arranged. In those days programme did not begin until 9.30am, but he took the idea to Alan James who

Pat (seated) with a group of his colleagues at the IBA, discussing engineering announcements. Left to right: Lynn Vale, Pat, John Brodzky, G3HQX and Peter Ashforth.

Pat in his office at the IBA.

became very enthusiastic. He discovered that once a week on Monday mornings the network came together to permit a play-out of 'newcomers' – the week's new commercials. It seemed technically possible to insert a five-minute announcement immediately before these 'Monday Newcomers'. He obtained the agreement of the Head of IBA Advertising Control and began to search for some introductory music that was out of copyright and did not breach Performer's Rights, leaving it to Pat to shape and script the announcements. At Brompton Road there was no colour camera, but the Experimental Department had a full-specification two-inch video tape recorder and a colour slide scanner. Peter King, the facilities manager, knocked up a black and white caption scanner that, with the aid of a Cox Box could put colour onto black-and-white 'captions' made up from white letters with magnetic backing stuck on black metal plates. It sounds crazy, but Pat says it all worked with EIS members operating the scanners so that captions and slides appeared alternately.

A search for announcers turned up two EIS secretaries with attractive voices – Lynn Vale and Anne Parsons. On Friday afternoons the crew would assemble in the ground-floor technical area and record the main announcements in readiness for Monday morning transmissions, with time left for a last minute live announcement of any engineering work planned for the week (usually involving transmitters running at half-power and subject to interruptions), details of which would often be phoned in at the last moment. It was just as well that none of them belonged to any of the powerful broadcasting unions, that would have been horrified to see who was doing what. Pat recalls one Monday morning when the 'announcer' arrived late and he took over the live spot.

After a time, Lynn Vale left the IBA and the search was on again for a new presenter. The choice fell on Peter Ashforth, who had recently joined EIS on gaining his second university degree and spending a year in Uganda. The choice was not unanimous because of his Yorkshire accent, but Pat convinced the critics that this was an advantage rather than a drawback. Peter was responsible for one of the few jokes when he once announced that a Welsh transmitter, Arfon, would be 'Arf off'.

The recital of transmitter information soon seemed a bit dull, so to liven things up Pat suggested that he and Peter should make unscripted recordings on general broadcast engineering topics that might interest the trade, having in mind the recordings he had made for the BBC World Service. This was done fortnightly and went down well.

Later in 1973, when the IBA Engineering Division moved to Crawley Court, the weekly Trade Announcements were transmitted from an Outside Broadcast-type vehicle in the basement garage of Brompton Road, with Pat and Peter (who had by then transferred to become an IBA Programme Control Officer) and the one remaining facilities engineer working the controls, with the audio material sent up the line from Winchester. The fortnightly interviews were still made at Brompton Road, using a Uher portable tape recorder. A problem was to find a place in Brompton Road that did not produce a noticeable reverberation. Eventually one was found – a stationary cupboard.

This continued until a small TV studio was finished at Winchester, after which the videotapes were made there with the interview audio recordings sent down the line to Winchester for inclusion in the VTR tape, later returned up the line again to Brompton Road for feeding into the ITV network. With the arrival of breakfast television (TV-AM) in the early 1980s, the Announcements were transferred to the new Channel 4 network, early in the morning (about 6am), with the intention that dealers would set their VCRs to record them.

Programme production was increasingly transferred to Crawley Court and increasingly with the occasional availability of a camera became a more polished affair. The 'programme' finally ended in 1989, when Pat made a final trip to Winchester to take part in a final recording – and a champagne celebration

of a 'programme' that had run for more than 20 years.

Digging-in the heels

No more then 18 months after joining the IBA there was once again a lot of talk about the Authority looking for a place in the country to relocate all their engineers to. Indeed, in 1973, this is exactly what came to pass. The IBA moved the whole of their engineering division to Crawley Court, a brand new building build on the site of a formerly magnificent – but by then derelict – country house that had stood near Winchester in Hampshire. Along with others in the department, Pat was asked if he would be prepared to move to Winchester or accept redundancy. Pat was among a small minority who said 'no' to the move and 'yes' to severance.

Because they wanted to keep him, Howard Steele and Alan James decided that Pat could stay in London to represent Engineering Information. Originally Alan James had arranged for John Brodzky, G3HQX, to stay at Knightsbridge along with Pat, but this was stopped because the management could see themselves ending up with a second engineering department in London. In the end Pat was the only member of the Engineering Division other than a facilities engineer and the 'Lines Booking' women (who were nominally in the Division) to stay at Knightsbridge.

A Bit of Controversy

One of the last editions of *Engineering News* that Pat produced for the IBA during his final years with the Authority and for a few months after he retired.

By that time Alan James had moved to another department and Dr Boris Townsend had become head of the Engineering Information Service. He had been brought in by Howard Steel from being Head of Technical Research at Thames Television. By the time the move to Crawley Court took place there were around ten people in the department, as it had grown by approximately two people per year. Pat much preferred working in London, plus it had the added bonus of his being left to his own resources.

During his first four to five years at the IBA, Pat was not only continuing to write 'TT', but also edited the Royal Television Society's journal *Television* and continued contributing his column to *Wireless World* etc.

It was in the early 1970s that Pat began to use factory-built equipment for amateur radio. First it was an old Hammarlund

HQ129x communications receiver, then a Labgear LG300 CW transmitter with a home-built 1.2kV power supply. Later still, a KW2000A transceiver, which is now on its last legs but still in use. Even with the transceiver CW remained Pat's favourite mode, although in recent years he has also used it for SSB, mainly on 3.5MHz nets.

Pat operating GB3WW in 1971, with a KW2000.

The shotgun wedding, called off

During the time that Pat was working for the IBA, a Government Committee was looking into the future of television broadcasting. One of the possibilities was that the IBA and the BBC engineering departments should merge, or at least the engineering research departments should. The IBA produced a submission for the committee, two chapters of which were written by Pat, who believes that it was partly attributable to the persuasive way he wrote his part that such a merger never took place. Pat wasn't in favour of competition for competition's sake, but he was most definitely in favour of a dual approach. With co-sited UHF transmitters, close co-operation on service area planning and good liaison between EIS and BBC's EID was necessary.

A familiar name of someone else who worked for a time in the Engineering Information Service of IBA is Rob Mannion, G3XFD, who later became the editor of *Practical Wireless*. Rob didn't work directly for Pat, indeed one of Pat's claims to fame is that he has never given an 'order' to anyone.

When he became editor of *Electronics Weekly*, Roger Woolnough wanted Pat to become their chief correspondent because he considered him the only person that could really be relied upon to be around at any given time and meet the deadlines. This offer was declined, as being responsible for about five extremely independently minded people wasn't Pat's idea of fun.

By the time Pat retired in 1987 he was a Principal Engineering Information Officer (Civil Service ranking AP4). Dated June of that year, the edition of *Engineering News* shown opposite contains product news, Papers, lists of meetings and conferences,

A Bit of Controversy

Front cover of the book that his IBA colleagues produced for Pat when he retired.

etc. It also contains a feature on electrostatic discharge lifted from *QST*, showing that Pat was causing information to flow back and forth between publications for professionals and amateurs. Significantly, the edition also contained a paper on 'A Proposal For a New High-Definition NTSC Broadcast Protocol' and a paragraph on HDTV which went as follows:

"At the height of the PAL/NTSC/SECAM controversy in the 1960s at least one major organisation issued a firm edict that its engineers should not attempt to develop their own new colour-encoding systems but instead should concentrate on solving problems on one of the three proposed systems which themselves were showing an alarming tendency to proliferate new versions. Judging by the number of systems now emerging, perhaps it is time that a similar ban should be placed on new systems for wide-screen HDTV".

History seems to be repeating itself, because just as this book was being written Toshiba announced that their HD DVD system for recording HDTV was being dropped because Sony's rival BluRay had taken the lion's share of the market.

Technical Topics

Pat considered 'TT' a sideline – certainly for many years, although it became more central to what he did after his retirement from the IBA in 1987. "For many years it's just one of those things one did", he says, adding "though I always enjoyed writing for radio amateurs."

The beginnings of the column

Good things don't just happen by accident, but having said that it needs to be said is that 'TT' was never actually commissioned by the RSGB.

In 1958 Pat was still a member of the Society's Technical and Publications Committee. Although no longer working for the society full time, early in 1958 he attended a meeting of the committee at which the *Bulletin* was discussed. The feeling was that, compared to *Short Wave Magazine* and *Radio Constructor*, the RSGB's journal was a bit dull. Various suggestions were made during the meeting, but no decisions made. Afterwards John Rouse, who was then the deputy editor, Roy Stevens, G2BVN, and Pat adjourned to a local pub to discuss what more could be done. One of the ideas put forward was to use snippets from other magazines as a regular feature, as opposed to a one-off article. Pat remembered that before the war Ken Jowers, G5ZJ, who was the editor of the short wave side of *Television and Short Wave World* used to publish a 'digest' page showing extracts from other magazines, including circuit diagrams accompanied by a small amount of text. Pat thought this was quite

Technical Topics

A survey of recent Amateur Radio developments—the first of a new regular series

By PAT HAWKER (G3VA)

TO keep abreast of current technical progress and practice in the Amateur Radio field has never been an easy task. New ideas and circuits are constantly being introduced and old ones revived. Some have a short life, others are absorbed into the main stream of amateur practice. Yet often, unless one has read the original article in a British or overseas magazine, it may be many months before one meets someone able to pass along sufficient details to find out what the latest technical trend may be, and to make it possible to try a new aerial or circuit device which may be just what the station needs.

We cannot promise that this new BULLETIN feature will solve all these difficulties. All we can hope to do is to survey from time to time a few ideas from the Amateur Radio press of the world; a few hints and tips that have come to our notice; with perhaps an occasional comment thrown in for good measure.

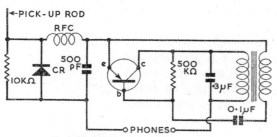

Fig. I. Simple transistorized c.w. monitor.

Transistors

It is almost impossible to open a magazine these days without finding some new or improved application of transistors. Yet many of us still tend to think of these devices as a rather expensive means of developing a few milliwatts of power on low frequency bands. That may have been true a year or two ago but already the power junction transistor has arrived. A single power transistor mounted on a small chassis as a heat sink can provide an audio output of 3-5 watts from a 12 volt supply. These transistors are suitable for modulators for mobile or low power work and also as oscillators in power supplies which offer many advantages over conventional vibrator supplies. Power transistors are currently available from several British manufacturers, though the prices are above those now ruling in the United States. There has been a spate of interesting articles in recent American and Continental magazines including: "Transistorized Power Supply" (*QST*, February 1958) describing a unit capable of delivering up to 65 watts at about 400 volts from 12 volt car batteries; an article on transistors in mobile work in the German *DL-QTC* (December 1957) providing details of a 10 watt h.t. unit and a 3 watt and a 1 watt modulator; a "Transistorized Meter Sensitizer" (*QST*, November 1957) showing how almost any low power junction transistor can be used in a simple circuit as a d.c. amplifier to convert a 1mA meter into an instrument with a full scale deflection of 50 or 100 μA, and a completely transistorized 10 watt modulator in *QST* (December 1957).

For the complete newcomer to transistors, a practical c.w. monitor was described recently in *DL-QTC* comprising a tone oscillator (almost any available transistor would be suitable) powered by the d.c. voltage output from an untuned crystal diode: see Fig. 1. One word of caution: make sure that the crystal diode is not radiating harmonics.

Neutralising Tetrodes

We were interested to see in *CQ* (Jan.-Feb. 1958) an article by W5OSL on bridge neutralizing tetrodes, as a similar circuit has been used at G3VA for some time. The basic circuit is now fairly well-known (although it has not been used in any BULLETIN equipment): see Fig. 2.

This is a bridge circuit which when balanced will effectively isolate the high impedance grid circuit from positive feedback due to grid-to-anode and stray capacitances. This is true when the ratio of NC : C1 is equal to the ratio of grid-to-anode capacitance : grid-to-cathode capacitance (i.e., $NC/C1 = Cg-a/Cg-k$). In most designs C1 is fixed but W5OSL makes the point (and this has been confirmed) that it is often much easier and safer to adjust the circuit by making C1 variable (about 500pF maximum).

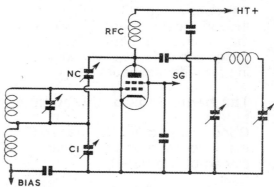

Fig. 2. Basic bridge neutralizing circuit for tetrodes This is suitable for almost any of the popular types.

Receiver Trends

In *QST*, January 1957 W1DX drew attention to the advantage of single conversion receivers with an i.f. above 1·8 Mc/s, provided that sufficient selectivity could be attained in the i.f. amplifier. A year later (*QST*, February 1958) W6YBR has come up with an inexpensive i.f. amplifier for c.w. using three cascaded 1690 kc/s surplus crystals but throwing out all i.f. transformers. This experimental amplifier has a bandwidth of about 1000 c/s at 20db down, and offers plenty of scope for further developments.

For those who want to try their hand at a fairly simple but practical receiver (at least for c.w.) there are many attractions in the "Super-gainer c.w. receiver" described by W6AJF in the bumper November 1957 issue of *CQ*. This is the latest version of an extremely simple design of some 20 years standing. The circuit uses a 6AB4 grounded-grid r.f. amplifier (which also functions as an automatic transmit-receive switch), a low noise 12AT7 mixer/oscillator, crystal filter and 6BJ6 i.f., and a $\frac{1}{2} \times$ 12AX7 as a regenerative second detector, the other half being an a.f. amplifier, with a 6A6 output. Regenerative second detectors are not particularly popular in this country but having used one for many years the writer can testify that when the regeneration control is smooth they can be highly practical for c.w. A small neon bulb is used as a voltage regulator.

The surplus price of FL-8 filters is likely to shoot up again if the " SAF-4 " QRM eliminator of W7OXD catches on (also in *CQ*, November 1957). This uses no less than four of these filters in cascade in an eight-stage audio filter, using four 6SL7 double triodes. This circuit gives a bandwidth of 160 c/s at 6db down and only 400 c/s at 60db down— figures which put normal 470 kc/s crystal filters or even 50 kc/s i.f. strips into the shade. But we suspect that it would lead to some hurried knob twisting between " overs " to try to find where some of those DX signals had drifted to!

New life for war-time receivers is the order of the day. In that connection W6SAI gives information (*CQ*, November 1957) on reducing the noise figure for the popular NC-240D by replacing the 6SK7 r.f. amplifier with a 6SG7 which has the same connections but a noise resistance only about one-third that of the 6SK7. The 6K8 mixer is replaced by a 6SB7Y (the equivalent noise resistance is about one-quarter that of the 6K8 but base connections differ). W6SAI also advocates the fitting of a voltage regulator.

Briefs

One big advantage of the folded dipole over the single wire type is that a quick and easy check for feeder and dipole breakages can be made by checking for continuity at the transmitter end of the feeder.

In a recent *CQ* survey, U.S.A. activity was found to be about 5 per cent c.w. only; 20 per cent phone only; 38 per cent mixed, with c.w. most of the time; 62 per cent mixed with phone most of the time. Ten per cent were using s.s.b. and 2 per cent operating radio teletype.

*　　　*　　　*

Please note that it is impossible for the author to loan or supply copies of the articles referred to in this feature. (Many of them have only been borrowed!)

Previous page and left: The first ever 'Technical Topics'.

a useful page and remembers that while he was working at RSGB in the early 1950s, a member offered a monthly feature along these lines. Entitled 'Bright Ideas' the first instalment was published and was quite good. Both the author and Pat understood that it was going to be a monthly feature, but for some reason it only appeared once.

When 'TT' first appeared in April 1958, its beginnings were humble. As you will see from the original 'TT' reproduced here in full size, it wasn't Pat's idea to solicit letters from readers, this only came later as a result of people writing to him. 1¼ pages and about 1,200 words in length, it equates to approximately one page in A4 format.

In the first 'TT' Pat wrote, under the sub-heading 'Transistors': "It is almost impossible to open a magazine these days without finding some new or improved application of transistors. Yet many of us still tend to think of these devices as a rather expensive means of developing a few milliwatts of power on low frequency bands. That may have been true a year or two ago but already the [AF] power junction transistor has arrived. A single power transistor mounted on a small chassis as a heat sink can provide an audio output of 3-5 watts from a 12-volt supply. These transistors are suitable for modulators for mobile or low power work and also as oscillators in power supplies which offer many advantages over conventional vibrator supplies. Power transistors are currently available from several British manufacturers, though the prices are above those now ruling in the United States. There has been a spate of interesting articles in recent American and Continental magazines...."

It was not the original intention to harvest information from professional journals for 'TT', but this may be because – except for *Electrical and Radio Trader* – Pat didn't have access to magazines of that kind while working at George Newnes. When he moved to *Electronics Weekly* the situation was completely different, because the publishers of *EW* received quite a lot of magazines and Pat was a regular user of the technical libraries.

For 'TT', John Rouse and Hutch would forward magazines to Pat, but the Society received amateur radio magazine from all over the world and not all were in English. Eventually, David Erwin, G4LQI, took over the non-English magazines and for several years used them to write his 'Eurotek' column for *RadCom*, because he was multi-lingual.

Original ethos

'TT' was originally envisaged as a monthly feature. John Clarricoats put the first offering in the April 1958 issue of the *RSGB Bulletin*. A second instalment followed in the May edition. The June instalment didn't appear in June, but in July. Clarry wanted 'TT' as a bi-monthly column after that, which wasn't Pat's intention at all. Nevertheless it remained as a bi-monthly column for ten years.

The April 1958 'TT' was one and a quarter pages. The May 'TT' was one page. Over the years it gradually built up. The largest 'TT' ever was twelve pages. This was caused by a strike, which had forced the Society to publish a double issue of *RadCom*.

Pat recalls visiting the newspaper printing works in South London that the Society was using at one time. Whilst there, he accidentally dropped a piece of paper. Naturally enough he bent down to pick it up. The chap who was showing him around immediately became very excited and distressed, and told Pat not to pick it up because it wasn't his job! Another man had to come over and do that. "Talk about Spanish Practices", says Pat, adding, "It was fatal". The number of strikes these days in the printing industry is nowhere near as many as in the 1950s.

In the immediate post war period there was a paper shortage. Along with many other commodities, it was rationed by the government. In those days the *RSGB Bulletin* was limited to 16 pages. Eventually the printers found – or diverted – paper from somewhere else. Because it came from a different budget, it didn't come out of the paper ration.

Pat got along very well with Hutch, who worked largely from home until the last few years of his editorship. In fact it was Pat who introduced Hutch into the RSGB in 1972, because the Society lost the previous editor suddenly. At the time there were two young chaps looking after the *RSGB Bulletin*. One, who was quite a reasonable editor, left and the other who was very young and had no experience except as an assistant editor took over for a couple for months before deciding that it was all too much for him. He emigrated to Australia, leaving the RSGB with no editor for several months. Roy Stevens, G2BVN, stepped in and got Pat and George Jessop, G6JP to help. Pat distinctly remembers going along to the then Society HQ at 35 Doughty Street (near Gray's Inn, in Central London) to help put the magazine together. At that time Pat was still going on press visits for the IBA and RTSJ. He remembers speaking to Bob Raggatt at one of them. He told Bob that RSGB were looking for an editor. Bob replied that he knew someone working for the Marconi Company who would quite like to leave. The name was given and duly passed on to the RSGB. He was called, invited to an interview and got the job, so he was always grateful to Pat. Because he wasn't a radio amateur and didn't fully understand the hobby, Pat used to read all the page proofs, mark up any corrections and send them back, but he would also proof read the final pages of the entire magazine. Pat is quite proud of the fact that he once prevented a potential libel getting through.

Producing copy for 'TT'

When Pat started 'TT' he wrote everything in longhand, using a ballpoint pen. He would then type it on his manual typewriter. Gradually the manual typewriter gave way to an electric model, but never to an electronic one. Whilst working at the IBA there was a short period when Pat could even get his secretary to type-up some of his copy.

Up until the mid 1990s Pat actually preferred writing his copy longhand, but then he bought a second hand Amstrad word processor from another amateur without realising that it was already obsolete. It was the Amstrad that ended Pat's column for *Wireless World*, because they wanted all copy electronically but couldn't handle Amstrad's file format. Pat still has the Amstrad machine, which works. Eventually Pat's son persuaded him to change over to an IBM-type PC running Windows 95, but this is now quite old and he is concerned that he will soon fill the hard drive.

In the early days of 'TT', producing illustrations for *RadCom* was all done by hand. There wasn't even any photo-copying. These days Pat photocopies a lot. For a long time illustrations would be sent to Clarry or John Rouse, G2AHL, but when Derek Cole took over as the technical illustrator Pat would send them direct to him instead. This was an arrangement that worked quite well. Derek and Pat would co-operate over corrections to the diagrams, so by the time the editor of *RadCom* saw them they were (hopefully) perfect. Without a scanner, Pat still sends hand drawn illustrations or annotated photocopies.

How 'TT' is produced these days

When Pat's material for 'TT' is received in the *RadCom* office, the text comes as a Microsoft Word file on a floppy disc. These days the normal size of 'TT' is four A4 pages, which equates to about 4,000 words. The words are taken by the editor and sub-edited to remove any typos, obvious errors and amend anything where the style is not quite right. There isn't usually much to do.

The illustrations that Pat sends in to accompany his text are invariably a mixture of annotated photocopies and hand-drawn sketches. These get passed to the illustrator/designer, who uses Adobe Illustrator to re-draw them.

When the text has been sub edited and the illustrations pro-duced, everything is printed out and sent to Pat by post for checking. When he has read it and added any corrections, he posts it back.

While the proofs are out, the pages will be designed. This involves 'flowing' the text into a desktop publishing package and placing the images and captions. These days the software used is Quark Express.

Readers who have not experienced professional publish-ing won't know that this process can be a little tricky, because there are certain criteria that must be met and pitfalls that must be avoided at all cost.

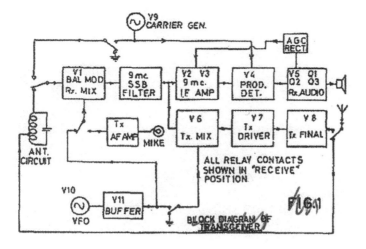

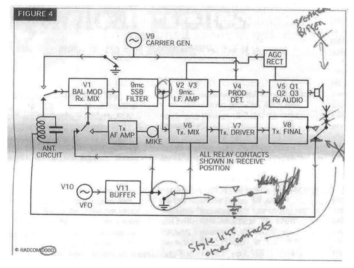

FIGURE 4

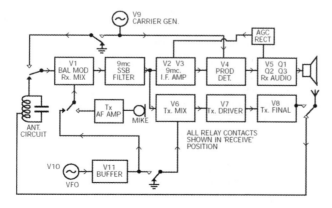

How a typical illustration for 'TT' evolves.

Top: Original, submitted as a photocopy.

Middle: The proof returned by Pat, complete with amendments and corrections.

Bottom: The final illustration as it appeared in *RadCom*.

A Bit of Controversy

The first criterion that must be met is that the copy must fill the pages exactly. There must be no overrun or blank lines left at the end of the final page. The trick is to adjust what is known as the 'tracking' of the text. Modern DTP software can adjust the gaps between characters and words in minute increments, which over the course of several pages can effectively shorten or lengthen a feature by a few lines. Individual words, lines and paragraphs can also be adjusted, to avoid placing the last word of a paragraph on a separate line or – the Cardinal sin – a new column.

Something that also needs to be avoided is the so-called 'bad break', where a title appears at or near the foot of a column of text. There are various ways to fix this problem. One is to adjust the size of the boxes that illustrations are placed in, forcing the text around them in a different way. Alternatively, an illustration can be moved elsewhere on a page. Finally, whole sections of text and the associated illustrations can be juggled. For that reason, the order of items when they appear in 'TT' is very often not the same as they were when submitted.

When the corrected proofs are returned, the changes are incorporated and the pages finalised, but is this really the end of the process? Not entirely, because there's no telling when 'TT' is received in the *RadCom* office if there's going to be too little or too much, or that it is possible to make all the pieces of the jigsaw fit together. What sometimes happens is that there is too much, in which case whole items need to be cut. These are not disposed of. Instead they end up as 'overmatter', which is kept for use at a later date.

In fact the very nature of 'TT' makes it relatively easy to assemble, because items of various lengths are submitted and it is not critical in what order they appear. If, for example, Pat has submitted an item that is 10 column inches long, but there is only 5 column inches available, a look through the overmatter will often turn up a held-over item that is close enough to the right size that it can be made to fit. The longer one then goes into overmatter, for use later. Sometimes it gets to the stage where there is a lot of overmatter. On the occasions that happens, Pat can simply be asked to submit less words one month, so the amount can be reduced. On the other hand, if there's too little copy, Pat always comes up trumps with a bit extra.

Having said that, Pat hasn't spent his life in publishing without learning about the requirement of writing to size, so more often than not everything does indeed fit.

New technology in 'TT'

Pat would regularly read *Nature*, which he sees as the leading science journal. *New Scientist* is also useful, but Pat can't get about like he used to, so visits to even his local library to use their photocopier or reference facility aren't easy. For the amount of copy he would get from *Nature*, subscribing himself is just not worth it. "*Nature* magazine is where you really see the first sign of anything new. The American magazine *Science* is also very useful."

Digital systems

Although Pat wrote a feature for *Electronics World* about the digial MP3 code, he admits to not knowing a great deal about digital electronics. He has a digital television, but only has a rough idea of how the digital part of it works. For producing copy for *RadCom* he has "this darned computer". All he knows is that every now and again it gives him problems because it does something he doesn't expect. He then has to press all sorts of buttons to try and find out how it works and why it did it. Having said that, he does understand how microprocessors work.

Basically, if its digital, it's difficult, but if Pat sees a straight-forward feature in a magazine he will start to study it. If it looks like a 'pie in the sky' concept or something he has read numer-ous times before, he'll pass over it.

Damp squibs

Pat remembers reading about cold fusion experiments at the University of Southampton and being very taken with the idea. He doesn't remember writing about it though, and it was even-tually found to be completely false. He cannot remember writ-ing about much that was later found to be completely false.

Pat was quite interested in – and read a lot of technical arti-cles on – telepathy. The trouble is, nobody has ever proved it happens.

Proudest moments

Being an extremely modest man, any mention of pride is quite alien to Pat. He is however very pleased that the Huff and Puff stabiliser was found interesting and that it caught on. Many regu-lar readers of 'TT' will remember it well, but what they won't know is that Pat was the one who named it Huff and Puff. It had appeared in a Dutch journal before 'TT'. With modifications, some people are still using it.

Greyline propagation is an aspect of amateur radio that found its way into 'TT'. Pat calls it the 'chordal hop' and attributes his knowledge of it to Les Moxon, G6XN. Les used to write a lot of long letters to Pat at one time, before - like so many of his early regular correspondents - going to that 'great shack in the sky'.

As far as Pat knows, one of the things he did write about early on is public key cryptography. He believes he was the first person to write about it in a British magazine (*Wireless World*). Even before WWII, Pat was reading books in his local library on codes and ciphers. Later, during the War, when he was in SIS, he used the Syko system to communicate with the 'secret navy' ships, 'poem' cyphers and one-time pads with the SIS code book, between SIS 'stations'.

With 'TT', Pat regards himself as a reporter, observer, digester and abstractor, but certainly not a designer. Pat feels that one of the reasons 'TT' caught on well is that most readers don't want to know in detail how something was designed mathematically, they just want to know how it works. Having said that, Pat himself is interested in how and when things were developed or invented.

Almighty arguments

Pat can't think of much he ever put in 'TT' that he later wished he hadn't, but there were certainly some things included that led to trouble. Features he wrote for other magazines also got him into hot water on a few occasions.

In one instance, both he and the Society were threatened with a libel action in the courts. Racal threatened to sue, because Marconi had published a feature in their in-house magazine that gave specifications for commercially available receivers. Marconi had named their own models, but just gave others models names like 'Receiver J'. Looking through the table, Pat could easily recognise what some of them were, because the IF frequencies were stated. Pat wrote in 'TT' that the specification given in the Marconi journal for the Racal receiver "was not terribly good". Racal were "terribly upset", threatened to sue Marconi for libel and the RSGB for repeating it. In the end Marconi admitted that they had made a mistake in the specification they had published for the Racal receiver and re-published a revised table with corrected figures. After the dust had settled Racal admitted to the then editor of *RadCom* that they were very pleased Pat had published the information, because that would have been evidence of which receivers Marconi were "cracking at". Pat still regards what he did was perfectly justifiable.

A large British telecommunications company was incensed with one thing Pat wrote and threatened to withdraw its representative from the Society's EMC committee. The bone of contention was cordless telephones and how some models were more susceptible to interference than others. It seems clear that an employee of the company who was also a member of the Society thought there was information in a 'guidance to service engineers' document that ought to be in the public domain, so sent it to Pat anonymously in a brown paper envelope. The trouble was that the company considered it confidential information, although Pat still feels that amateurs were entitled to see this guidance material.

A magazine once got very upset when Pat reproduced a valve table from their publication in 'TT'. Pat's response was along the lines of, 'Why do you keep sending me complimentary copies of your journal if you don't want me to repeat anything from it?' The storm blew over.

The longest-lasting technical dispute is over small loop antennas. The main devotee still maintains he is right, whereas most other people say he is wrong. There is no question that the radical theory was not produced in good faith; Pat simply says that he has over-estimated the radiating efficiency.

Multi-path reception of FM and how useful it is for broadcasters to use circularly polarised aerials to reduce this became a real hot potato. Unfortunately it was just at the point when the BBC was about to change over from linear polarisation to circular at some of their transmitting stations. Pat wrote a feature in *Wireless World* saying that its effectiveness was open to question. The BBC Director of Engineering wrote a bitter letter of complaint to the Director of Engineering at the IBA. He took it up with Pat, but *Wireless World* had not only credited the feature to Pat, they had put his IBA affiliation against it. This upset Pat's boss no end, and apparently he made enquiries into whether he could get Pat fired for having done it. In the end Pat was seconded from the Engineering Division of the Authority to the Information Division, but he continued writing his articles.

Another 'spot of bother' was in *Television* (RTSJ), where Pat wrote a very critical review of a book that had been written about John Logie Baird. In Pat's opinion the book contained a number of outrageous claims about what Baird had done in his lifetime. Pat has a certain amount of admiration for the man, but believes he also had his down sides. It caused another strong letter to be written, this time to the Chairman of the IBA, complaining about the review. The IBA passed it to Tom Robson, the Director of Engineering at Crawley Court, who passed it to Dr Boris Townsend. Townsend wrote a conciliatory reply for Robson to sign, which he did and the letter got sent. The trouble was that this was done without Pat's knowledge, which of course annoyed him. The argument continued for quite some time, so if you want Dr Waddell to go ballistic you only need to mention Pat's name in his presence. The feeling is mutual though.

Frustrations

Pat seriously considered packing 'TT' in a couple of times. At around the 30th anniversary of the column it was cut back from six pages to four pages per month. However, Pat acknowledges the fact that the page size and consequently the publication space per page of *RadCom* has increased over the years, so the current four pages is approximately equivalent to six pages in the old magazine format.

Whenever a new *RadCom* editor came along, changes occurred. Sometimes it might be to put all the references at the end of the article, whereas Pat always likes to put references in the text because he believes that is where people see them. So-called 'white space' on pages, which has come and gone in *RadCom*, caused Pat some irritation. Pat admits that he tends to over-write and fondly remembers 'Hutch', the editor who would simply find more space in the magazine to put 'TT', with a "continued on page..." message at the end of where 'TT' should have finished. In those days it wasn't necessary to write to an exact length, but things are different now.

Copyright issues

There have been virtually no difficulties in copyright over the years, and no major ones, although Pat's original intention when 'TT' started was only to quote amateur radio journals, not professional ones. Abstracting only some words from an original article and quoting the full source and author usually forestalls any problems.

A mistake?

It is questionable that a 'TT' item on 'a very short dipole', along with a claim that it worked as well as a full sized dipole, should have appeared. It was taken from the IEEE Transactions on Antennas and Propagation. Being a professional body, the IEEE is generally regarded as completely trustworthy, but *RadCom* readers tried to replicate the design and then started to report back

Pat, giving a talk to a radio club in the 1980s.

to Pat that they couldn't get anything like the results that were claimed for it. In the end it was accepted that 'a very short dipole' was pretty hopeless, because it just didn't have the radiation efficiency that was claimed.

For his part, Pat would claim that publishing this item was not a mistake, because by encouraging a couple of American amateurs to replicate the design it was possible to challenge the original academic author and set the record straight.

Official Recognition

Pat has a 'fan club' and because the last thing you would ever describe him as is an attention-seeker, any mention of it makes him chuckle. Over the years numerous people have corresponded with him on technical matters and countless numbers have experimented with ideas published in 'TT'. The secretary

A Bit of Controversy

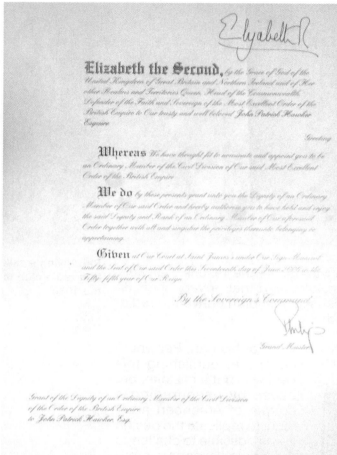

Pat's MBE certificate, signed by Her Majesty The Queen. Awarded for Services to Radio Communications.

of many a radio club has written to Pat, asking him to attend and give a talk. This he used to do frequently, and although he found public speaking difficult the members invariably showed their appreciation with a round of applause. Many *RadCom* readers have written letters to let him know how much they enjoy reading his column and to thank him for his sterling efforts. Some of them decided to write to Buckingham Palace and nominate Pat for an award.

The British honours system has been around in one form or another since the fourteenth century. These days there are ten awards and one appointment available for people who have performed outstandingly or given service of merit. Nominating someone is relatively straightforward, although not all nominations lead to an award. The MBE (Member of the British Empire) is the 'base' award in the line that includes the OBE, CBE and Knight/Dame. It is awarded for one of two things: Achievement or service in and to the community of a responsible kind that is outstanding in its field; or very local 'hands-on' service which stands out as an example to others.

Pat received his MBE from His Royal Highness the Prince of Wales at Buckingham Palace on 14 December 2006. Recipients are allowed to take three guests with them to the awards ceremony, and you

THE MOST EXCELLENT ORDER OF THE BRITISH EMPIRE

O.B.E. & M.B.E. (Ladies in uniform, and Gentlemen)

The Insignia should be worn in the following manner:-

The Badge, suspended from its ribbon, should be worn on the left side by gentlemen, and by ladies in uniform in accordance with the Dress Regulations of the Service concerned.

When evening dress is worn with Decorations, the Badge should be worn in miniature.

The Insignia is not returnable on death, but should be retained by the person legally entitled to receive it under the terms of the deceased's Will. Notification of the date of death of the recipient should be sent to the Secretary, Central Chancery of the Orders of Knighthood, St. James's Palace, London SW1A 1BH.

SEE OVER

even get to park your car at the Palace while it takes place. Invariably there are a number of awards presented per ceremony, and Pat remembers there must have been about 110 recipients there on that day. The way it works is that the guests are all seated in a room, which is set out in the style of a theatre. All the recipients wait in an adjoining room. When their name is called they walk into the main room and are presented to the member of the Royal family who is presenting the awards that day. After the investiture and an exchange of pleasantries, the recipient goes off to collect his/her medal case, then joins his/her guests. The next recipient is then called in.

Something Buckingham Palace never tell a recipient is who nominated him/her, so although Pat has a pretty good idea, he cannot be certain.

Half a century

Pat retired from writing 'TT' on a monthly basis when it reached its 50th anniversary in April 2008. He says that he could easily have stopped a few years ago, when an extremely critical letter about his column was published on the letters page of *RadCom*, but chose instead to see it out to its 50th anniversary. Pat now finds typing and getting about difficult, but accepts that people might still want to write to him. Those who choose to do so can be assured that their input will not find its way straight into the waste paper bin. Instead, they can expect to read 'Technical Topics Extra' in a future *RadCom*.

Selected Bibliography

Early Articles
'An Interesting Hobby' – (Short-wave-listening). *Minehead County School Magazine*, July 1936.
'Television at Radiolympia and the Science Museum, 1937'. *Minehead County School Magazine*, July 1938.
'G3VA'. *Minehead County School Magazine*, July 1937.
The above three articles were republished in *Radio Bygones*, February/March 2002 under the title: 'As it seemed then ... a youthful look at the 1930s'
'Wireless Direction Finding' by "Navigator". *Armchair Science*. Reprinted as 'Tell-Tale Radio – How the BBC Baffles the Nazis' in *The English Digest*, 1941.
'The Voice of a Great Democracy - World-wide Transmissions, Flash News and Vital Information to Remote Places' (Rugby Radio GBR) - an unsigned article in *Modern Wonder*, 1941.
'The Future of Amateur Radio' by "Navigator". *Wireless World*.

February 1941.
'Statistics for the Radio Amateur'. *RSGB Bulletin*. 1941.

RSGB Publications
A Guide to Amateur Radio. Contributor to 6th edition, edited by John Clarricoats, G6CL. Editor of 7th to 19th editions (1958-1983). Hardback versions of 17th and 18th editions published by Butterworths, by arrangement with the RSGB.
Technical Topics for the Radio Amateur. 1st Edition 1965, then title changed to progressively enlarged editions of *Amateur Radio Techniques*. 2nd edition, 1968. 3rd edition, 1970. 4th edition, 1972. 5th edition, 1974, (reprinted) 1978. 6th edition, 1978, and final 7th edition 1980 (reprinted 1991).
Technical Topics Scrapbook. Series of pages reproduced from 'Technical Topics'. 1985-89, published 1993, 1990-1994 in 1999, 1995-1999 in 2002, 2000-2004 in 2005, 2005-2008 in 2008 (plus 50 years of 'TT' CD-ROM).
Antenna Topics. Cuttings of antenna items from 'Technical Topics' 1958-1999. Published 2002.
The Amateur Radio Handbook. Contributor of 'HF Receivers' chapter of 3rd edition, 1961. Title then changed to *Radio Communication Handbook*. Contributor of 'HF Receivers' chapter of 4th edition, 1968; 5th edition, 1976; 6th edition, 1994.

George Newnes Technical Books
Radio Engineers' Servicing Manual. Published under the names Molloy & Poole, 1953.
Radio & Television Servicing. 1st edition (2 volumes, 1953) with annual editions up to 1965 (early editions under the names Molloy and Poole, later editions under the name J P Hawker, or Hawker and Reddihough).
Radio & Television Engineers' Reference Book. 1st edition, 1954; and 2nd edition, 1958. Published under the names Molloy & Pannett. 3rd & 4th editions published under the name J P Hawker & Pannett.
Television Engineers' Pocket Book, with J A Reddihough and specialist contributors. 1st edition, 1954. 2nd edition, 1958. 3rd edition, 1960. 4th edition, 1962. 5th edition. 6th and subsequent editions edited by J McGoldrick, etc.
Radio Servicing Pocket Book. 1st edition 1955 published under the names Molloy & Hawker. 2nd edition 1962, under name J P Hawker.
High Fidelity Sound Reproduction (with specialist contributors). 1st edition published under the name E Molloy, 2nd edition as Molloy and Hawker.
Electronics Pocket Book (with J A Reddihough). 1st edition, 1963. 2nd edition, 1967.

Radio and Television Reference Data. Compiled by JPH, 1963.
Outline of Radio & Television, by J P Hawker, 1966. Published in the USA as *Radio & Television: Principles and Applications* (Hart-Publishing Company).

IBA Publications
Engineering for Colour. ITA, 1969.
Contribution to: 'Evidence to the Committee on the Future of Broadcasting' (under the chairmanship of Lord Annan). Chapter 6, 'Engineering'. Chapter 7, 'Continuity and Change'. IBA, 1974.
Technical Editor and Contributor to *IBA Technical Review No 3 – Digital Television*, 1973.
Technical Editor and Contributor to *Television Transmitting Stations*, IBA Technical Review No.4, 1974.
Technical Editor, *IBA Technical Review No.8 – Digital Video Processing – DICE*, 1976.
Technical Editor and Contributor to *IBA Technical Review No 11 – Satellites for Broadcasting*, 1978.
Technical Editor and Contributor to *IBA Technical Review No 17 – Developments in Radio-frequency Techniques*, 1982.
Broadcasting Technology for the 1980s. A 12-page brochure, with foreword by Tom Robson, IBA Director of Engineering. 1980.
Editor, *IBA Engineering News* (monthly). Contributor, 1962-1984 (approx). Editor and compiler, 1985-1987 (approx).

Contributions to Books
The Secret Wireless War by Geoffrey Pidgeon, UPSO, 2003. Chapter 34 'Pat Hawker – His Many Roles in the Secret Wireless War'. 'Appendix 5 – Agents Sets' and sundry other items'.
Electronics Engineer's Reference Book, edited by F F Mazda (Butterworths, 6th edition, 1989). Television.
Telecommunications Engineer's Reference Book, edited by Fraidoon Mazdam.1st edition, 1993, and 2nd edition. Butterworth/Heinemann. Television.
The Electronic Epoch, by Elizabeth Anteébi. Published by Van Nostrand Reinhold Company, 1982. Diodes.

Oral/Paper Presentations
'Keep it Simple: Direct-conversion HF Receivers'. IERE Conference Proceedings No.40, Southampton, July 1978.
'Effect of Receiver Specifications on Receiver Performance'. IERE Clerk Maxwell Commemorative Conference on Radio Receivers and Associated Systems, Conference Proceedings No.50, Leeds, July 1981.
'Satellites and Cable Television'. Presentation to The Society of

Cable Engineers at Savoy Place on 16 October 1979. Published (with discussion) in *Cable Television Engineering*, February 1980. 'The Early Days of Clandestine Radio'. Colloquium on 'The History of Military Communications, Proceedings Fifth Annual Colloquium", Centre for the History of Defence Electronics, Bournemouth, September 1999.

Evening Lecture, 'Clandestine Radio in World War II'. IEE, Savoy Place, and repeated to IEE Section at Swindon, 1994 (two-hour audio tape of lecture and questions held by Pat).

Regular columns

'Technical Topics'. *RSGB Bulletin/Radio Communication/ RadCom*, April 1958 - April 2008 (bi-monthly 1958-1968).

'World of Amateur Radio'. One-page (later expanded to four pages) as 'Communications Commentary', 'Broadcast', 'RF Connections', etc. 1969-1992.

'Radio World'. One-page column for *Everyday Electronics*, 1980-1985.

'The Pat Hawker Column'. *Viewpoint*, 1980-91.

Miscellaneous articles, including Clandestine Radio

'Fifty Years for the advancement of Amateur Radio - A Survey of Amateur Radio over the years and the part played by the RSGB'. *RSGB Bulletin*, July 1963. Also reprinted as a 12-page brochure.

'Bulletin Reflections'. Marking the 50th anniversary issue of *Radio Communication*, July 1975.

'Synchronous Detection in Radio Reception'. *Wireless World*. Part 1, September 1972; Part 2, November 1972.

'Using Directional MF Transmitting Antennas'. *Communication and Broadcasting*, Summer 1976.

'Times Remembered: 21 Years of TV Engineering'. *Independent Broadcasting*, September 1976. An extract from this article appears in Chapter 35 'The Technical Background' of Volume 2 - 'Expansion and Change' of the official series *Independent Television in Britain* by Bernard Sendall (Macmillan Press Ltd, 1989).

'News – The Electronic Revolution'. *The BKSTS Journal*, July 1977.

'Ten Years of Colour - 1: Choice of Colour System'. *Television* (IPC), December 1977.

'The Broadcasting Scene – A Year of Technical Questions'. *Educational Broadcasting International*, December 1977.

Low-cost Satellite Receiving Systems. *Wireless World*, January 1978.

'Teletext, Available on All Four Television Networks in The UK, is Established, Accepted and Growing'. *Television/Radio Age* (US),

4 July 1983.

'Goodbye to 405 lines'. *Television* (IPC). January 1985.

'50 Years of TV: The Beginnings'. *Television* (IPC). November 1986.

'Montreux Show to Focus on 'New Techniques' as Exhibitors Plan 625-line Product Launchings'. *Television/Radio Age* (US) 23 May 1983.

'90 years: Serving Amateur Radio'. *RadCom*. RSGB 90th anniversary issue. Part 1, July 2003; Part 2, August 2003.

The Early Years of the Amateur Service' (a postscript to the 90-years two-part article). *RadCom*, July 2005.

'60 years – A D-Day to Remember'. *RadCom*, June 2004.

'50 Years Have Slipped Away'. 50th anniversary issue of *OTNews*, journal of the Radio Amateur Old Timers' Association, April 2008.

Five-part series in *D-I-Y Radio* (RSGB). Part 2, 'Radio: A Unique and Exciting Hobby', September/October 1993. Part 2, 'The Dawning of International DX', November/December 1993. Part 3, 'Radio in Peace and War', January/February 1994. Part 4, 'Amateurs as Secret Listeners', March/April 1994. Part 5, 'A Unique Hobby for Life', May/June 1994.

'Amateur Radio Developments'. *Wireless World*, November 1951.

'Amateur Radio Progress'. *Wireless World*, November 1960.

'Battle of the Circuits – Cables Take on Satellites in Telecommunications and the Honours are Even', unsigned, *British Industry Week*, 8 December 1967, and reprinted in *Zodiac* (Cable & Wireless journal).

'Small talk of the Micro-circuit Men', unsigned, *British Industry Week*, 28 March 1968.

'Weather Forecasts – Outlook Brighter'. *British Industry Week*, 1 November 1968.

'The World of Amateur Radio, 1911-1971'. 60th anniversary issue of *Wireless World*, April 1971.

'A D Blumlein – Father of Modern Stereo', by Patrick Halliday, *Electronics World* (US) about 1960.

'British Audio Trends', by Patrick Halliday. *Electronics World* (US), 1961.

'Clandestine Radio in WW2'. *Wireless World*. Part 1, January 1981. Part 2, February 1981.

'The Secrets of Wartime Radio'. *Amateur Radio* (Goodhead Publications), March 1983.

'Electronics News Gathering'. *Television* (IPC). August 1976.

'The Outlook for AM Stereo'. *Radio Month*, 20 February 1978.

'Preparing for the Fourth Channel Transmissions'. *The BKSTS Journal*, June 1980.

The Pioneers of Television - From Concept to Post-war Reality'. *Electronics Australia*, September 1980.

A Bit of Controversy

'Television Coverage – The Next Steps'. *The BKSTS Journal*, November 1980.

'Gerald Marcuse, G2NM'. *Radio Bygones*, October/November 1989.

'The Hague Concerts from PCGG – The Bitter Sweet Story of Henso Idzerda'. *Radio Bygones*, April/May 1990.

'The Mysterious A J Alan – Leslie Harrison Lambert, G2ST'. *Radio Bygones*, June/July 2001.

Electronic Cryptography – Codes, Ciphers, Communications and Computers. *Wireless World*, September 1980 (included the first simplified description of 'Public key' systems in the UK, plus some account of WW2 cryptography and cryptanalysis, etc).

'The Pioneers of Television. *Television* (JRTS), November/December 1983.

'Bugs & Sideswipers on the Clandestine Links'. *Morsum Magnificat* No.14, Winter 1989.

'P P Eckersley [G2OO] – The Missed Opportunities'. *Radio Bygones*, February/March 1992.

'Clandestine Radio'. *Morsum Magnificat*. Part 1, No.22, 1992; Part 2, No.23, 1992.

'The First Time I Saw Paris'. *Morsum Magnificat*. April 1994.

'The Polish Radio Centre'. *Radio Bygones*, June/July 1992.

'The Biggest Aspidistra in the World'. *Radio Bygones*, August/September 1992.

'International Broadcasting – Birth of a Monster?'. *Radio Bygones*. Part 1, 'Looking Back on Short-wave Listening in the 1930s', June/July 1993. Part 2, 'Broadcasting in the 1930s and 40s', August/September 1993.

'Aspi 5, Task Z and Operation 'Silent Minute' – The Secret Story of the Vain Attempt to Jam the V2'. *Radio Bygones*. Part 1, August/September 1994. Part 2, October/November 1994.

'The UK/PRC-316 (A-16) and Early Solid-state Manpacks'. *Radio Bygones*. February/March 1997.

'Origins of the El-Bug'. *Morsum Magnificat*. August 1997.

'Nuvistors, Compactrons, Frame-grid and Quick-heat Valves'. *Radio Bygones*. June/July 1997.

'FM Radio Faces the Digital Threat'. *Radio Bygones*. Part 1, 'Developments up to WWII', Christmas 1997. Part 2, 'Return to Peace', February/March 1998.

'Agenten Funker im Welt Krieg – Beobechtungen Zeitzeutgen' (translated into German and published by the Editor). *Amateurfunk Jahrbuch*, 1993.

'Birth of Broadcasting'. *Electronics World*. May 1997.

'Talking Dangerously ... Delving Deeper- [S-phone, Ascension, Rebecca/Eureka]'. *Practical Wireless*, January 2002.

'Harold Kenworthy, OBE, [G6HX] – His Key Role in Signals Intelligence'. *OTNews* No.61, Winter 2002.

'Plan Sussex - 1944'. *Radio Bygones*. Part 1, 'Preparations', October/November 2004. Part 2, 'Operations', Christmas 2004.

'German Amateurs in WW2'. *Radio Bygones*. Part 1, 'Amateur Operation and Equipment', February/March 2006. Part 2, April/May 2006.

"Czech Clandestine Radio from the UK". *Radio Bygones*. Part 1, 'Flight to the UK and Agent A-54', October/November 2006. Part 2, 'Assassination and its Aftermath', Christmas 2006.

'Covert Radio in Holland, 1944-45'. *Radio Bygones*. Part 1, 'Market Garden and its Aftermath', February/March 2007. Part 2, Triumphs and Disasters, April/May 2007.

'50 years have slipped away'. Golden Jubilee edition of *OTNews*, April 2008.

Profiles

Radio Personality Series: 'Pat Hawker, G3VA' as told to Rob Mannion, Editor, *Practical Wireless*, September 1993.

Profile Series: 'Pat Hawker, G3VA' as told to Ken Hill, G3CSY, for *Mercury* - the Royal Signals Amateur Radio Society Journal. Part 1, No.129, July 2001. Part 2, 2001.

Miscellaneous

'The Colour Television Controversy', by Patrick Halliday. *Practical Electronics*. February 1965.

'Why Intelsat'. *Zodiac*. 1968.

'An Introduction to Integrated Circuits and Digital Electronics'. *The SERT Journal*. February 1974.

'TV Receiving Aerials'. *Television* (IPC). March 1976.

'Citizens' Band – The Pros and Cons'. *Practical Electronics*, September 1976.

'CB and the Radio Amateur'. *CB World*, 1970.

'Independent Local Radio in the UK: Broadcasting and Communication'. Marconi publication, 1974.

'The Use of Computers in Independent Broadcasting'. *National Electronics Review*, January-February 1976.

'The Story Behind Teletext', by Patrick Halliday. *Television & Home Video*, July 1979.

'Digital Video – How Far has it Got?' *Broadcasting Systems & Operation*. July 1979.

'The Impact of International Technical Conferences – IBC90'. *Combroad*. September 1980.

'Radio & TV Servicing, Molloy & Camm'. *Radio Bygones*, February/March 2001. Recollections of working at Tower House, the production of the series of *Radio & Television Servicing* and how to identify the various editions).

'An Invention that Changed the World – The Birth of the Transistor'. *Vintage Wireless*. Part 1, Vol.19 No.1, 1993. Part 2, Vol.19,

No2, 1994.

Occasional articles for a wide range of broadcasting trade journals and general periodicals including: *The Guardian* (newspaper), *Electronic World* (US), *Television/Radio Age International* (US) and also their London 'Connections' market reports, COI (Central Office of Information), *IEE News*, *EBU Technical Review*, *ASBU Technical Review*, *Television* (IPC), *Television* (RTSD), *IBE* (International Broadcast Engineer). Incidentally, no attempt has been made to list the mass of material in *Electronics Weekly* or in-house IBA publications, book reviews, etc.

Television appearances

ITV News at Ten: Interviewed on Russian spy trawlers, 1968.

East of England/BBC2: Participation (but no on-screen appearance, except for a photograph) in final 30-minute version of 'Secret Listeners'. Researched by Paul Cort-Wright, G3SEM, 1977. BBC2, 1979.

BBC2: 'World at Their Fingertips'. RSGB 'Open Door' programme for BBC2. February 1979, presented by Brian Rix, G2DQU. Scripted item on HF propagation.

ITV (Granada): World in Action. Interviewed on Russian electronic spying in the USA, 1980s.

ITV (Granada). 'All Change – First Memories'. Programme about early computers, Alan Turin, Bletchley Park and Manchester University. Early 1980s.

BBC1: '9 o'clock News'. Interviewed by Laurie Margolis, G3UML, about the ending of Maritime Morse. Late 1990s.

IBA 'Engineering Information for the Television Trade'. Passim 1972-1989. Mostly as audio recordings with added slides.

Radio Broadcasts

Number of broadcasts (including phone-ins) on Independent Local Radio Stations about reception problems. Also LBC on Oracle, teletext, etc. A few on BBC London etc. BBC World Service 'World Radio Club' about 1967-1988. Passim at first, later infrequent.

'TT' Supplement

The final regular 'Technical Topics' to appear took the form of a supplement that was inserted into the April 2008 edition in *RadCom*. In it Pat looked back at the birth of 'TT', thanked some of the many people who contributed to it, revisited some of the topics covered, lists some of the sources he used and thanked those who conveyed their appreciation of his efforts. It is reproduced in its entirety on the following pages, reduced slightly in size from the original format.

Bulletin reflections

The Techical Topics supplement is followed immediately by 'Bulletin Reflections', a feature Pat wrote in July 1975, when the RSGB's journal – by now renamed *Radio Communication* – celebrated 50 years of production. These pages are reproduced in their original size.

TECHNICAL TOPICS

50 Years in retrospect – Pat Hawker looks back over the half century of developments covered by TT

NNIVERSARY THOUGHTS. For this 50th Anniversary *Technical Topics"* it seems appropriate to look back and call how the column came into existence, its original objectives, how it developed, a few of the many diverse topics and innovations it introduced, and then to contemplate if only briefly what the future may hold for amateur radio.

OW DID *TT* BEGIN? In 1958 I was a member of the Society's "Technical & Publications Committee" (T&P) that met quite often at the then HQ – the 4th floor of 28 Little Russell Street, WC1, near the British Museum. In 1958, the membership had fallen from its high peak of the immediate post-war period, largely due to the problem of TVI. There was also some criticism by members of the monthly *RSGB Bulletin*, popularly known as *"The Bull"*, with some members comparing it unfavourably with *"Short Wave Magazine"* edited with a professional touch by Austin Forsyth, G6FO.

The *Bull* of the 1950s had some excellent innovational and constructional articles but by 1958 seemed to be running out of steam. With the exception of the HF, VHF, SSB and Mobile columns, articles were subject to a slow-moving peer-review process and were expected to reflect only work based on the practical experience of the author.

There was even a suggestion that the Society should approach G6FO and see if *SWM* could become the Society's journal. At the T&P Committee meeting there was discussion on how the *Bull* could be made to appeal to more members, but nothing firm emerged. So after the meeting, Roy Stevens, G2BVN (an active Council Member), John Rouise, G2AHL (Deputy Editor to John Clarricoats, G6CL) and I adjourned to a nearby hostelry to chew over the problem and try to think up some practical ways of improving the appeal of the *Bull*.

At the back of my mind was a monthly one-page feature in the pre-war "Television and Short Wave World" compiled by its Short Wave Editor, Ken Jowers, G5ZJ comprising abstracts (including diagrams) of articles published in overseas magazines. Again, during the years (1947-51) that I had been a staff member of HQ (finally as Assistant Editor), I had abstracted a few single pieces from overseas journals: for example 'Simple CW/Phone monitor' from *Radio Revista*; "Parallel Cathode Modulation" from *Amateur Radio (WIA)*; 'A Variable Frequency Crystal Oscillator" from *Amateur Radio (WIA)* (the first amateur description for what is now known as a VXO); and the 'cascode' low-noise amplifier from the original paper in *'Proc. IRE'* and since widely used in the pre-amplifier stages of HF and VHF receivers. There had also been an attractive once-only feature "Bright Ideas" (by a contributor who soon fell out with G6CL) that had been intended as the first of a regular series but never reappeared.

I tentatively suggested that I should try my hand at a new monthly column based on digests and short hints and tips. G2BVN and G2AHL welcomed my suggestion. I duly wrote to Clarry sending him a draft of what became the first *"Technical Topics"* of April 1958, sub-headed "A survey of recent Amateur Radio development – the first of a new regular series". It was by-lined "By Pat Hawker (G3VA)", the first time I had used the name 'Pat' in print. For previous published material in the *Bull* and elsewhere I had used my initials 'J. P.' or a nom de plume – a deliberate attempt to lighten the rather formal approach then standard practice in the *Bull*.

Not being a professional radio engineer I realised that I was taking the risk of making technical bloomers, particularly when abstracting material from foreign-language texts. I have to confess that once, in the early days, I took a circuit diagram of an "audio-filter" from *'Radio-REF'* that turned out to be an April fool joke. The filter placed a virtual short-circuit across the audio! A few of the antennas abstracted from professional as well as amateur journals have made exaggerated claims. I have also learned to be careful in using the word "first" which seems inevitably to remind one or more readers of some long-forgotten but basically similar idea. This is but one reason why it is worth reading old as well as current magazines and books, and taking an interest in how practices evolved, only to be overlooked; yet remain well worth reviving. I remain in awe at the ingenuity and inventiveness of the pioneers who worked with spark or the early valves. There is a tendency for each new generation of amateurs to disregard the continued value of past work.

In that first *'TT'* I set out my stall as follows: "To keep abreast of current technical progress and practice in the amateur radio field has never been an easy task. New ideas and circuits are constantly being introduced and old ones revived. Some have a short life, others are absorbed into the main stream of amateur practice. Yet often, unless one has read the original article in a British or overseas journal, it may be many months before one meets someone able to pass along sufficient details to find out

what the latest technical trend may be, and to make it possible to try a new aerial or circuit device which may b just what the station needs.

"We cannot promise that this new *Bulletin* feature will solve all these difficulties. All we can hope to do is to survey from time to time a few ideas from the Amateur Radio press of the world; a few hints and tips that have come to our notice; with perhaps an occasional commen thrown in for good measure."

While these aims have been adhered to over the years they have been extended in two significant ways. After a few issues, comments and suggestions began to come in from UK and overseas members/readers; in the second place I began to find information in the professional journals that had not previously appeared in RSGB publications yet seemed highly relevant to amateurs. So increasingly *'TT'* became recognised by members and eve some non-Amateur professional engineers as a forum providing early publication for technical ideas and new developments.

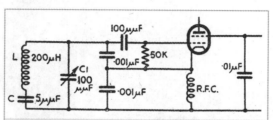

FIGURE 1: A real 3.5MHz Clapp variable frequency oscillator described by G3PL in 1949 in the RSGB Bulletin. Note the recommended values for C an L (see text). Bandspread is about 100kHz when C is about 5pF.

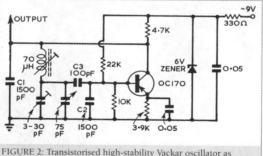

FIGURE 2: Transistorised high-stability Vackar oscillator as developed by BRS25769.

SOURCES. Many of the *'TT'* items have come from such sources as *QST, CQ, 73 Magazin Electronics, Electronics World (USA, G-E Ham Notes, RCA Ham Tips, Electronics Australia, Australian EEB, CQ-DL, Radio-REF, Radio-ZS, DL-QTC, Sprat, UBA, Electro Break-in, Interadio – 4U1ITU, Radio (Moscow), Wireless World (later Electronics & Wireless World), Electronic Engineering, Television, EBU Technical Review, Signal, Electronics Weekly* (on which I worked from 1963 to 1968), *Ham Radio, Communications Quarterly, QEX*, etc, and house journals such as *Point-to-Point Communication, Marconi Review*, etc. Institutional journals have included *Proc IRE* until it merged with *Proc IEEE*, IEEE's *'Transactions Antenn & Propagation'*, IEEE's *'Antennas and Propagation Magazine'*, IERE and IEE journals, IEE's *'Electronics Letters', Proc IREE Australia*, etc. Book sources have included ARRL publications; *"The Radio Handbook"*; some of the books which I compiled or edited while with George Newnes Technical Books (1952-63), or from technical publication of the ITA/IBA's Engineering Information Service where I worked from 1968 to 1987, or the Conference Books of the International Broadcasting Conferences that I attended in Montreux, London and Brighton and the National Broadcasting Conventions I attended in Chicago Washington DC and Dallas. Not all these sources now urvive or remain unchanged.

I have always tried when abstracting material to
rovide full reference and acknowledgement to the
riginal author and publication and have found that in
9.9% of cases both author and journal have been pleased
 have their work recognised in *RadCom*. There was one
reat of a trade libel, but it was withdrawn.

Although most early digests were from overseas
mateur radio publications, I found that technical
evelopments reported in the professional journals, trade
nd house magazines, available to me at libraries or at
ork, included interesting material of potential interest to
mateurs. Again, after joining *"Electronics Weekly"* in early
963, I attended many IEE and IERE meetings, seminars
nd conferences as well as
ress visits to UK and overseas
rms, research establishments
nd international conferences.
his continued after I joined
he Independent Television
ater Broadcasting) Authority
TA/IBA) in October 1968,
ecoming also for some years a
are-time Editor of *"The Royal
elevision Society Journal"*. Many
ooks and journals were
anned at The Science
luseum Library (now largely
erged with an Imperial
ollege Library); the Patent
ffice Library (now part of the
ritish Library); the IEE
ibrary (now IET Library); the
3A Library (now dispersed).
nfortunately most of these
braries have seen changes
lat either make them less
seful as *'TT'* sources or in the
ast three years less reachable
ue to my arthritis. And
acreasingly RF analogue HF
nd VHF engineering material
 now swamped by digital
aformation technology based
n software.

All these sources provided
T' useful information on new
evelopments and new
omponents during the years when much professional as
ell as amateur effort was still being put into HF and VHF
adio communications equipment, antennas and
ropagation, using analogue and subsequently digital
ystems. Much of the professional HF research and
evelopment has of late been cut back, affected by the
nding of the Cold War, the development of global
atellite communications, short-range mobile phones, the

internet and allied information technology. The London
libraries have ceased holding on their shelves copies of
many of the amateur radio-based journals that they
formerly held. Years ago the IERE merged with the IEE;
the IEE is now part of the IET and for some years have
held fewer evening meetings and largely abandoned their
one-day seminars at Savoy Place. The British radio
communications industry has changed profoundly with
the disappearance of such major firms as Plessey, AEI,
Marconi, EMI Electronics, Racal, Redifon, BCC, GEC and
so on.

The BBC and IBA no longer own the UK radio and
television transmitters, with the IBA becoming first ITC
and now part of OFCOM. The Government hopes that
broadcasting will eventually be
based solely on the complex digital
systems and allow it to sell off
much of the spectrum, although I
suspect that AM and FM radio will
not disappear for many years. The
public is accepting DTV and
anticipating that HDTV will
flourish. But DAB and DRM are
struggling – and HF SSB
broadcasting which should have
been mandatory by now has
virtually disappeared off the radar.
The growth of cable distribution
has been slowed by terrestrial and
satellite Freeview. Broadcasting
seems to be seen increasingly as
just part of IT, distributed in digital
form over the web or by satellite to
laptops, i-Pods, mobile phones etc.

All these changes have made the
compilation of *'TT'* more difficult
and more time-consuming; indeed
the need for it is more questionable
in these days of factory-built
equipment and with so much
information available on the web.

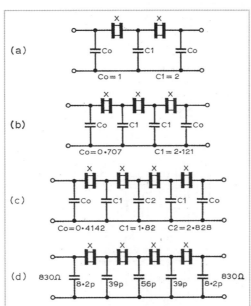

FIGURE 3: Early crystal ladder filters investigated by F6BQP.
All crystals (X) are of the same resonant frequency and
preferably between 8 and 10MHz. To calculate values for the
capacitors multiply the coefficients given above by 1/(2Pi f
R), where F is the crystal frequency in Hertz, R is the input
and output termination impedance and 2Pi is roughly 6.28.
(a) Two crystal unit with relatively poor shape factor. (B)
Three crystal filter can give good results. (d) Four crystal
unit capable of excellent results. (d) Practical realisation of
four-crystal unit using 8314kHz crystals, 10% preferred value
capacitors and termination impedance of 820 ohms. Note
that for crystals between 8 and 10MHz termination
impedance should be between about 800 and 1000 ohms for
SSB. At lower crystal frequencies use higher design
impedances to obtain sufficient SSB bandwidth.

THANKS. I had intended at this
point to list and thank some of the
many who have contributed novel
circuits, antennas, oscillators, filters
(crystal, mechanical, LC etc). But
there have been so many
contributors to *'TT'*, both UK and overseas – literally
hundreds – that it would be invidious to single out and
select what could only represent a limited selection.
Nevertheless such names as 'Dud' Charman, G6CJ (SK),
Les Moxon, G6XN (SK), Dick Rollema, PA0SE, Peter
Martinez, G3PLX, Klaas Spaargaren, PA0KSB (SK), Dave
Gordon-Smith, G3UUR, Jan-Martin Noeding, LA8SK
(SK), Brian Austin, G0GSF, Jack Belrose, VE2CV, Colin

Horrabin, G3SBI and Gian Moda, I7SWX spring to mind, but there are many, many others to whom 'TT' is indebted. It has always pleased me that 'TT' reaches and is read by so many overseas amateurs and professionals. Much of my time has been in dealing with and responding to the many letters and queries that have reached me. I take this opportunity to apologise to those readers who have submitted ideas that for one reason or another (including the occasional mislaying of them) have not appeared, and to some correspondents to whose letters, even after extensive searches, I have been unable to answer satisfactorily, or not at all.

So now, in this 50th Anniversary 'TT', just a tiny selection of some early items that stick in my memory as being developments that have (or should have) subsequently become accepted as part of standard practice; concentrating on topics that did not appear in the recent "50 year reviews" or are not otherwise available in the *"Technical Topics Scrapbooks"*.

CLAPP & VACKAR
OSCILLATORS. The immediate post-war decades saw intensive search for VFOs of improved stability. Voltage regulation, temperature compensation, high-Q coils, careful choice of components and their placement away from heat sources led to improvement of the classic Hartley, Colpitts and Franklin circuits. Then, in the late forties, J K Clapp described an oscillator based on a crystal-equivalent circuit (it was later shown that the BBC engineer, Geoffrey Gouriet, had developed and used professionally a similar arrangement during WW2). Amateurs soon adopted this oscillator although often failing to observe the correct choice of component values, even after A G Dunn, G3PL ("Clapp or Colpitts?", RSGB Bulletin, June, 1949) pointed out: "Many versions of the Clapp oscillator have been published in recent months, but it is evident that the main difference between the Clapp and Colpitts have escaped the attention of the designers of some of these so-called Clapp circuits. The majority are found on close inspection to be little, if-at-all, better than the original Colpitts circuit ....". After

discussion of what constitutes a crystal equivalent circuit he provided an example of a real Clapp VFO for 3.5MHz, **Figure 1**.

G3PL stressed the low value (5pF) of C and the high value (200µH) of L. He noted that C is often shown as a variable capacitor for tuning purposes but this is not essential. As a result the value has often been given as 40 or 500pF in some published circuits. L is of a value more usually associated with medium-wave circuits, and is some twenty times as large as the values commonly used in VFO circuits for 3.5MHz.

Perhaps it was the misuse of the Gouriet/Clapp circuit that led to further investigation of stable variable frequency oscillators. In 'TT' ("Transistorised Vackar Oscillator", July, 1966) I wrote "Thanks to L Williams, BRS25769, of Birmingham, we are able to include a circuit which appears to have real promise for both transmitters and receivers but which, as far as we are aware, is described here for the first time."

BRS25769 wrote: "I have done some work on a high stability oscillator circuit (**Figure 2**) which is a transistor version of the Vackar-Colpitts (Tesla) circuit. I am sure someone must have done it before but have never seen it in print. The circuit arose from a search for a VFO for a high stability receiver... Published circuits seem to be either of the Hartley or Clapp form, usually with buffer amplifiers. The Clapp has the feature of the output varying with frequency and transistor buffers do not give as high a degree of isolation as a well-designed valve stage.

"The Vackar circuit has the transistor terminals shunted with very large lumps of capacitance, as are also the output terminals. The values given cover 2.0 – 2.5MHz, and the prototype will stay zero beat with a crystal standard for hours. With C1 and C2 polystyrene and C3 silvered-mica, temperature drift is plus 10Hz per degree Centigrade; this could be improved by making C3 mixed mica and ceramic. The amplitude of oscillation is controlled by C2. Increasing C2 reduces amplitude without very much effect on frequency. For good stability

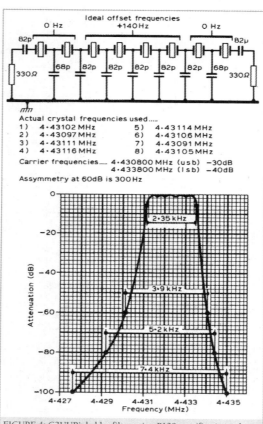

FIGURE 4: G3UUR's ladder filter using P129 specification colour TV 4.43MHz crystals showing the excellent shape factor etc that can be achieved.

08

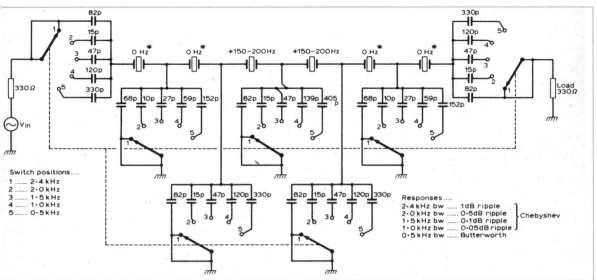

FIGURE 5: G3UUR's design for a switched variable-bandwidth filter using colour TV crystals. Note crystals shown as 0Hz offset can in practice be +/-0Hz without too detrimental effect on the passband ripple.

Switch positions....
1 .... 2·4 kHz
2 .... 2·0 kHz
3 .... 1·5 kHz
4 .... 1·0 kHz
5 .... 0·5 kHz

Responses ....
2·4 kHz bw ...... 1dB ripple
2·0 kHz bw ...... 0·5dB ripple
1·5 kHz bw ...... 0·1dB ripple
1·0 kHz bw ...... 0·05dB ripple
0·5 kHz bw ...... Butterworth
} Chebyshev

e amplitude [with an OC170 pnp device] should be mited to a few hundred millivolts. ..."

RYSTAL LADDER FILTERS. An item "Making Crystal dder filters!" ('TT', September 1976) noted that for ears most bandpass crystal filters had been based on the alf-lattice or lattice configuration: "Such filters require ie use of a number of crystals of carefully defined (and ifferent) frequencies, and often the use of centre-tapped iductive components - inevitably expensive items to buy id at HF very difficult to construct.

"Relatively little has been published in amateur ournals about an alternative filter configuration – the dder network – that appears to offer very useful features those who wish to save money by building their own lters. The little that has been published – for instance in irrent advertisements for the Atlas solid-state transceiver suggests that very high ultimate rejection figures can be chieved....

"The absence of information on building ladder-type rystal filters has to some extent been rectified by a most seful article by J Pochet, F6BQP in 'Radio-REF' (May, 976) covering filters using two, three and four crystals all f identical frequency: Figure 3.

"F6BQP shows clearly that very useful SSB filters can e made, seemingly with few problems, by anyone having n hand, say, four identical crystals. His prototype filters ere based on 8314kHz crystals ...but can be made at any equency from roughly 5 to 20MHz, although the apacitances and impedances favour the use of 8 to 0MHz."

The November, 1976 'TT' showed a CW ladder filter sing four 8794kHz and a 50-ohm input and output npedance, adding that J A Hardcastle, G3JIR was

preparing a detailed article following extensive work on the design and construction of ladder filters that had begun before my reporting that of F6BQP. His four-part article "Some experiments with high-frequency ladder crystal filters" ('Radio Communication' December, 1976, January, February and September, 1977) remains a classic. A shorter version was published in 'QST'.

For 'TT' (June 1977) Hans Kreuzer, DL1AN reported on ladder crystal filters based on the low-cost 4.33618MHz crystals used in colour television receivers to provide the colour sub-carrier reference frequency for the synchronous demodulation of colour signals. Since then PAL and NTSC (3.5796875MHz) crystals have been widely used for low-cost CW and SSB ladder filters. Years later, in April 1999, Jan-Martin Noeding, LA8AK showed that useful LF and MF ladder filters could be constructed using ceramic resonators.

G3UUR showed ('TT', December 1980) how ladder filters with excellent characteristics and/or switched bandwidths could be implemented, although requiring rather more careful selection of PAL crystals: Figures 4 & 5.

More complex than a ladder filter, attention was drawn in 'TT' (December 1969) to the continuously variable-bandwidth symmetrical IF filter used in the Rohde & Schwarz receiver Type EK07-80 using two high-slope low-pass filters and four mixers: Figure 6. Despite its exceptional performance, most readers seem to have decided it was too complex to copy. However, quite recently (see 'TT', July 2002 and October 2004) Dick Rollema, PA0SE reported his use of this approach (which he called a "sliding doors" system) with two 10kHz low-pass filters in a high-performance, home-built HF transceiver. Unlike the R/S filter, his could be adjusted for

09

LSB, USB or symmetrical response, each of continuously variable bandwidth: **Figure 7**.

COMPONENTS, REPAIRS & SAFETY. Over the years, *'TT'* has provided early information on many new components and active devices, plus a great number of hints and tips on checking, servicing and constructing equipment. This has included simple test instruments, making of and fault finding on printed circuit boards, soldering techniques and 'third-hand' devices, soldering jigs, handling surface-mount devices and components etc. Some of the ideas have been derived from trade and industry magazines or overseas amateur journals, many others been contributed by readers. Unfortunately, it is now impracticable to recommend home servicing the current generation of factory-built amateur transceivers with their surface mounted components, except possibly to those with a solid background of professional or extensive amateur experience.

However, older hard-wired valve or solid-state equipment based on discrete devices and/or boards using standard size ICs and discrete components can still, with care, be tackled at home with a limited number of test instruments and tools, with some experience. I have to confess that I have spent some months trying, so far unsuccessfully, to trace a fault on an ancient AR88 - the main problem is proving its sheer size and weight, leading to my virtual inability now to move it around!.

Although today one sometimes feels that there is too much attention paid to safety and too little to experimentation, *'TT'* has highlighted the potential hazards arising from electric shock, toxic chemicals in components, servicing aids, asthma inducing solder-flux fumes, and eye protection while soldering or metal working, cadmium in transistor casings etc, erecting and dismantling antennas and masts or using ladders; wartime luminous paint in surplus equipment; and some critical looks at the banning of lead in solder.

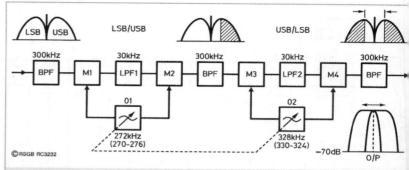

FIGURE 6: Basic principles of the 1969 R/S EK07-80 filter which uses two 30kHz low-pass filters in conjunction with ganged oscillators tuning in opposite directions. Variable from about +/-150Hz to about +/-6kHz with similar slope at all settings down to -70dB.

DIRECT-CONVERSION RECEIVERS. *'TT'* has always shown interest in receivers that can be easily constructed yet capable of providing practical CW and SSB reception. This may reflect my own use pre-war of a simple two-valve (0-v-1) receiver with regenerative detector and a transmitter limited by my licence to ten watts DC input which enabled me to work some DX including North & South America, Australia, and Southern Rhodesia (Zimbabwe). The regenerative detector can be extremely sensitive and fairly selective when just oscillating (good for CW and not too bad for SSB) but the gain and selectivity suffer badly when out-of-oscillation and used for AM or in the presence of strong AM signals. Simple regenerative and VHF super-regenerative receivers have been featured a number of times.

In March 1967 I expressed what has always been a tenet in compiling *'TT'*: "One way of developing something new is to look back at ideas that have been around for a long time but which have never been widely used. Often basic principles are developed many years in advance of the materials and devices that make them practical or economical. Poulsen's magnetic recording of 1889 had to wait until the 1940s brought forth the practical domestic tape recorder; Blumlein's stereo disc techniques of 1929-31 were not used until the past decad [1950s]; Robinson's 'stenode' tone correction with a single-crystal filter was largely forgotten until G6XN revived the idea in 1962".

As an example, in November 1967, I wrote: "In a recent *Electronics Weekly* we drew upon these histories to

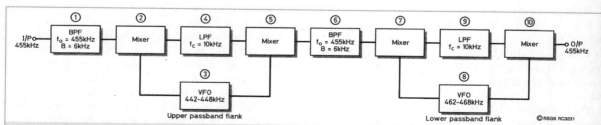

FIGURE 7: Block diagram of PA0SE's "sliding doors" filter.

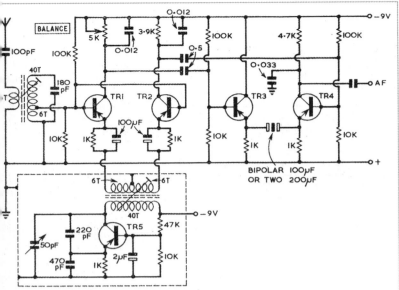

IGURE 8: Circuit diagram of PA0KSB's homodyne-type direct-conversion 3.5MHz receiver, first ublished in the Dutch journal 'Electron' and then in 'TT' (March 1967) was one of if not the earliest ansistorised receivers of this type Transistors were pnp AF124s in the original but almost any RF pes would be suitable (with npn types voltages should be reversed). The coils were wound on hilips-T formers.

uestion whether it is not time to look again at nchrodyne or homodyne receivers. The synchrodyne n be regarded either as a superhet with an IF of 0kHz or a straight [direct-conversion] receiver with a balanced near heterodyne detector. The true synchrodyne is the mplest form of phase-locked receiver, and in essence onsists simply of a balanced mixer (product detector), a cal oscillator locked to the incoming carrier, and an idio amplifier: no IF strip or second detector, no images r spurious responses, and – at least in theory – the ability govern selectivity entirely by the bandwidth of the igh-gain] AF circuits, a far cheaper way of achieving igh selectivity than with a crystal or mechanical filter.

"Of course, if it were all as simple as the last sentence uggests we would all have thrown away our superhets ears ago. We recall, back in the 'forties, in conjunction ith [Charles Bryant] GW3SB trying to get a Tucker nchrodyne circuit to work on AM broadcast stations ithout much success. But then, a lot more is known oday about linear balanced mixers, and about phase-ocking. And again, for SSB and CW reception there is no eed for the local oscillator to be phase-locked to the ncoming signal. Some time ago, 'TT' drew attention to /2WBI's 3.5/7MHz receiver of this type ('QST', May 961) using two well-balanced 6SB7Y valves as eterodyne detector, with a separate well-screened eterodyne oscillator – but the item did not attract much ttention. ...

"A positive sign that some amateurs are still onsidering the possibility of using such an arrangement

– this time with semiconductors – is to be found in the Dutch journal 'Electron' (January, 1967) where [Klaas Spaargaren] PA0KSB describes the front-end of a simple 3.5MHz receiver for CW or SSB: **Figure 8**. Two transistors form a balanced detector followed by a form of differential amplifier to produce an unbalanced AF output suitable for feeding into a conventional AF amplifier, with the fifth transistor used as a heterodyne oscillator. Using a 7m long aerial wire, PA0KSB says that he has been able to hear VE1 stations on 3.5MHz in the winter evenings…"

Quite soon, this form of receiver received a major boost with the publication in 'QST' November 1968 (summarised, in 'TT', February 1969) of an article "Direct Conversion – A neglected technique" by Wes Heyward, W7ZOI and Dick Bingham, W7WKR, with full constructional details of a 3.5MHz receiver capable of copying CW signals down to below the microvolt level and of sufficient stability to give reasonably good SSB performance: **Figure 9**."

Throughout the following years, this form of direct-conversion has been widely adopted, with many variations, as the basis for low-cost, home-built receivers, despite the presence of the audio image. Later on, they included the addition of image-rejection techniques using phasing-type filters (including those based on digital ICs, 'third-method' and polyphase networks). This approach, although adding complexity, opened the way for direct-conversion receivers with a performance as good as, or even better, than conventional superhet receivers, as found for example in the Elecraft K2 transceiver kit. Direct-conversion also forms the basis for some of the "software defined radios" currently attracting attention.

POLYPHASE FILTERS. While the vast majority of circuit developments reported in 'TT' have represented the work and ideas of amateurs and/or professional radio engineers, I can claim to have played a small part in the application of 'polyphase' filters to HF SSB transmitters and receivers. In October 1973 I wrote: "All amateur transmitters employ one of three systems for generating SSB: (a) the filter method; (b) phasing ('outphasing' or quadrature) method in analogue or digital form (**Figure 10**); and (c) the 'third method' (Weaver or Barber). .... It would be exciting to report that some entirely new method of SSB generation has been developed which combines all the

11

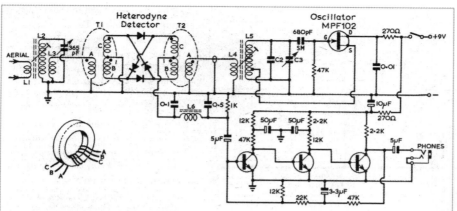

FIGURE 9: Circuit diagram of the 3.5 or 7MHz homodyne-type direct-conversion receiver by W7ZOI in 'QST' (November 1968) and in 'TT' (January 1969) that set the pattern for the home construction of simple but effective receivers. FET oscillator C2 is 470pF S.M. for 3.5MHz, 120pF S.M. for 7MHz. C4 is 140pf to cover 3.5 to 4.0MHz, 40pF for 7MHz. The three bipolar transistors were all RCA 40233 but other AF types could be used. L1, L3 3-turns, L2 40 turns No 28 enam. on 0.680-in toroid; L4 5 turn link; L5 22 turns No 28 enam. on 0.680 toroid with tap 5 turns from earthy end. L8 is 88mH toroid (a small smoothing choke might prove reasonably suitable). Since 1967 a large number of variations on this general type of receiver and hints on improvements etc have been published in 'TT' and RadCom.

virtues and none of the vices of all three systems, providing, say 60dB of sideband suppression at low cost and without unduly critical component values. It would be even more sensational to publish a circuit diagram showing exactly how to assemble such a system.

"I cannot do the second – but it is possible to draw attention to an article which explains the principles of a system which seems to fill most of the requirements: "Single-sideband modulation using sequence asymmetric polyphase networks" by M J Gingell, of STL, in *Electrical Communication*, Vol 48, No 1 – 2 (combined issue) 1973, pp21-25. Gingell's basic polyphase network is shown as in **Figure 11**.

"My difficulty is that even after reading it through several times and phoning the author, I find it difficult to attempt to translate a pretty involved and complex paper into terms which I, and I suspect many readers, would understand. ... It is claimed that a sideband suppression of 60dB can be achieved consistently in a polyphase network of capacitors and resistors with tolerances as much as +/- 2.5% at the input section to about +/- 0.2% at the output. In fact by adding or taking away extra sections one can design for varying degrees of performance and component tolerance. Up to 70dB of suppression has been achieved."

Michael Gingell's work at STL was in connection with line telecommunications at around 100kHz but in our telephone conversation he could see no reason why a polyphase network should not be used in HF transmitters. Years later, I learned that he had been disappointed that STC (ITT) had decided to use conventional line filters and his work had been shelved.

I concluded the item by writing: "I hope that by drawing attention to this novel technique I may stimulate somebody who really understands or can grasp how a

sequence asymmetrical filter really works and how it could be designed into an SSB generator suitable for amateurs to build. It would be possible to make a bit of communications history by doing so, since gather that no commercial applications have yet appeared. So how about somebody?".

Peter Martinez, G3PLX was quick to respond. The December, 1973 'TT' included an item "More of polyphase SSB" which began: "Peter Martinez, G3PLX, has done a valiant and valuable follow-up to the October item on the polyphase system for SSB generation. Although he has not had time to build a complete SSB generator he has gone back to the original paper by M J Gingell and translated the mathematics into practical terms, and has built several polyphase networks with results well in line with expectations: for example **Figure 12**.

He sees the polyphase network as, in effect, a new way of making a wide-band audio phase-shift network – but with the advantage of being much more tolerant of component values than the well-known approaches. With a six-stage network, which can be realised using readily available components, he believes that 10% tolerance can be used in the first five stages and two per cent in the final stage, yet this should be capable of providing some 40dB sideband suppression. .... To complete the SSB generator, Michael Gingell feeds the a, b, c, d outputs to four modulators, and combines the outputs; each modulator being fed with RF carriers in the sequence 0, 90, 180, 270 degrees. This has the advantage of cancelling out harmonic sidebands which occur on either side of the harmonics of the carrier frequency. But a simpler arrangement would be just to feed a and b to two balanced modulators fed with carriers at 0 and 90 degrees ..."

G3PLX provided a most useful simplified explanation of this type of network including the advantages of using a six-section network, permitting the use of readily available preferred-value components.

It was a pity that by the time that Gingell's polyphase system was published in *'TT'* the era of large scale home-building of SSB generators had subsided, with the result that it has not made a serious impact on amateur transmitter practice. However there is little doubt that it can be an effective and useful configuration, now cognised by both amateur and professional designers.

As I reported in TT, July 2001, Michael Gingell worked
n polyphase and related techniques for some ten years,
om about 1966, by the end of which time STL's interest
d switched to digital filter techniques. In 1974 he
ceived a PhD from London University for his work on
olyphase filters and SSB. He later moved to the States
d became KN4BS.

'TT' March and April 2001 included an account of
me subsequent developments in the use of polyphase
tworks in receivers, including the experimental design
a 14MHz high-performance, single-signal, direct-
nversion receiver by G3OGW using a four-phase
olyphase filter as an SSB demodulator together with the
e of FST3253 switching mixers. An outline of G3OGW's
perimental design is shown in **Figure 13**. Unfortunately
3OGW became a Silent Key before converting his
ototype into a fully operational multiband receiver – but
t before showing that this could become a valuable
proach still to be fully
ploited in receivers or
ansceivers. This is an area that
uld well form the basis for
rther experimentation,
ssibly linked with DSP and
her software defined
chniques.

NOMALOUS PROPAGATION.
'T' has always shown interest
reporting and commenting on
omalous propagation
nditions that can extend the
nge of LF, MF, HF, VHF and
icrowave signals. Such modes
ay permit communications
ove the MUF on HF or well
yond the horizon on VHF and
ove. Examples include
6XN's early work on extremely
w angles of elevation by which

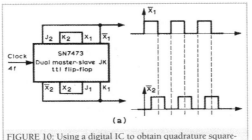

FIGURE 10: Using a digital IC to obtain quadrature square-wave signals.

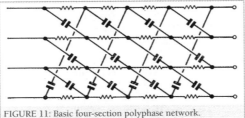

FIGURE 11: Basic four-section polyphase network.

e achieved SSB contacts with Australia using only 1 watt
om a sloping site adjacent to a salt-water loch; Albrecht's
hordal hop; whispering galleries; gray-lines; forward-
nd back-scatter, antipodal enhancement, round the
orld echoes, etc at HF and 50MHz.

Anomalous modes include Sporadic-E, meteor scatter,
oposcatter, ionospheric scatter, moonbounce, trans-
quatorial, rain scatter, "aircraft-enhanced" modes at VHF
nd above. Because many of these VHF modes are fleeting
r unpredictable they tend to be of more interest to
mateurs than to professional communicators, other than
roadcasters who often regard them as sources of
nwanted interference.

One of the more significant post-war discoveries in the
eld of HF propagation was that made by Dr H J Albrecht,

VK3AHH/DL3EC, when he realised that the signal
strength and reliability of European amateur signals
received in Australia on 3.5, 7 and 14MHz could not be
accounted for by the conventional theories of multihop
propagation, leading him to propose the theory of long-
distance 'chordal hop' HF propagation without
intermediate ground reflection losses. This concept,
together with similar work by Fenwick and Stein on
"around-the-world echoes" and the satellite-orientated
work on "whispering galleries" have received many
mentions in 'TT' and has gradually led some of us to
believe that the *majority* of long-distance HF contacts
made by amateurs using low power are made without
intermediate ground reflections rather than conventional
multi-hop mode. Sea reflections attenuate signals less
than ground reflections so that multi-hop paths are not
uncommon. Albrecht's work was brought to my notice in
the early 'fifties by Les Moxon, G6XN.

A 1979 article in 'Telecommunication Journal' by German
External Service broadcast
engineers showed in detail how a
superior service to Australia and
New Zealand could be provided
(Australian evenings, European
mornings) by utilising the
24,000km long path across South
America on about 8MHz rather
than via the 16,000km short path,
often up to 25dB stronger than
that calculated from CCIR
formulae for higher frequencies
nearer the MUF.

Less successful have been the
attempts to investigate and/or
explain the "long delay echoes"
(LDEs) that have been credibly if
rarely reported since the 1920s.

COMING UP TO DATE AND
THE FUTURE. Two relatively
recent developments that have
appeared first in 'TT' have been the wide dynamic range
H-mode mixer and. the low-phase noise twin-tank
oscillator, both representing significant state-of-the-art
circuit configurations. These were originally developed by
Colin Horrabin, G3SBI, in the mid-1990s but since
further improved with the aid of the latest IC devices.
Since these topics have been covered in recent issues and
in the series of 'Technical Topics Scrapbooks' they are noted
here only as examples of the continued role of amateur
radio developments contributed by those with both
professional and amateur experience based on analogue
and digital RF engineering.

It is clear that future developments will increasingly be
governed by improvements in IC devices in which more
and more bits work at increasing speed at ever lower

13

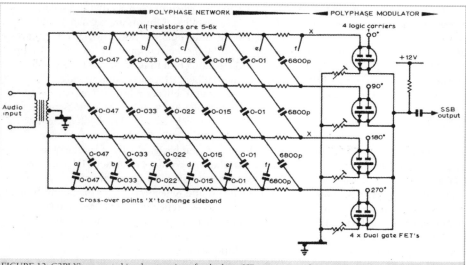

FIGURE 12: G3PLX's suggested implementation of polyphase SSB generator with six-section network using preferred values

and for the home constructor or inform[ed] critic of factory-built equipment stemming from an appreciation o[f] the role of the individual component[s] and their limitations. There are still controversies to settle[:] the emergencies of global warming seem likely to call for thin-line systems, includin[g] the use of NVIS HF propagation, supplied from improvised powe[r] sources. There will sti[ll] be enthusiasts for restoring and repairing vintage equipment. Nostalgia can be a powerful learning tool.

voltages. But even here there are many who believe that we are already approaching the limits imposed by silicon and even gallium arsenide, not to mention the problem that devices can demand all the skills of a watch-repairer equipped with powerful optical binocular magnifiers.

This perhaps is why many of those amateurs who believe that home construction should still form at least part of the hobby are turning back, as Peter Chadwick, G3RZP recently suggested, to the use of discrete components and to replicas or refurbishment of valve equipment. I make no apology for the space devoted since the 1970s to the clandestine radios of the Underground groups and Prisoner of War receivers. There is also still room for experimental work with crystal sets, as noted in *'TT'* as recently as January 2007.

Many years ago I wrote: "Digital techniques utilising microprocessor intelligence and either with or without substantial electronic memory can do many useful things, Such systems can sense, can control, can encode or decode, can convert one code into another, can change the rate of flow, can retrieve information in many different arrangements, can display information – all with accuracy, and unlimited endurance. Unlike the human operator, the microprocessor does not grow tired and careless, and once the program has been fully debugged does not itself introduce mistakes".

By now, we have learned (though not necessarily used) that effective filtering, mixing, demodulation can all be achieved by digital processing. With modes such as PSK51, signals can be received successfully from below the noise level. The human brain, however, can still hold its own in being more flexible and adaptable and amazingly good at solving those problems that depend on some form of pattern recognition. I believe, or at least hope, that there is still a role for the human operator –

Nor should we overlook or forget the pioneers of radi[o] and electronics. As someone born before the first transmissions of the British Broadcasting *Company*, I feel privileged that I have either met or attended lectures by such 'greats' as Dr Zworykin, Alec Reeves, Dr George Brown (RCA), Sir Robert Watson Watt, Sir Bernard Love[ll] Sir Martin Ryle, G3CY and may possibly have brushed shoulders at Hanslope Park with Alan Turing.

But I have to admit that, looking back for this Jubilee issue to the early days of compiling *'TT'* in my limited spare time, I cannot help feeling that the first 25 years were the most rewarding and personally satisfying, writing then to readers who had known the joys and frustrations of home construction, who saw that HF radi[o] communication still played a useful, unchallenged role i[n] global communications. Today, well, *sic transit gloria* – but let us believe with some certainty that there will in futur[e] still be a role for the technically-minded, experimentally-minded Radio Amateur!

ADDENDA. I had intended to end this special retrospective Anniversary *'TT'* at this point but there are still some matters outstanding, arising from recent issues that require clarification and/or amendment rather than being held over for a possible *"Technical Topics Extra"* in a few months time.

ANTENNA TRANSMISSION LINES. In the long leading item: "Is there nothing new in antennas" (December 2007, pp77-79), in the section on SWR, I inadvertently added in parenthesis some misleading advice. As printed (1st column, p79) the sentence read: "Note that when an antenna element presents to the transmission line an impedance other than its characteristic impedance, the

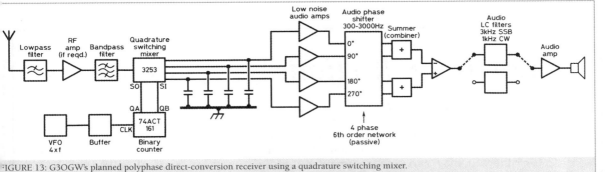

FIGURE 13: G3OGW's planned polyphase direct-conversion receiver using a quadrature switching mixer.

npedance offered to the transmitter at the input end of the line may be quite different from either the characteristic of the line (unless the line is an exact multiple of an electrical λ/2) or the impedance at the antenna junction."

The note (added as an afterthought) "(unless the line is an exact multiple of an electrical λ/2)" is incorrect. Stewart Rolfe, GW0ETF writes: "Forgive me if I am misinterpreting your sentence as published but it is my understanding that in such situations the impedance offered to the transmitter will 'never' equal the characteristic impedance of the line. At an electrical half wavelength from the antenna feedpoint, the (lossless) line impedance will be the same as that at the feedpoint itself, being in effect once round the line circle on a Smith Chart where the (normalised) impedance is the circle's centre point. The Smith Chart thus illustrates how the line characteristic impedance will only ever be presented on a flat line (ie no reflections or standing waves).

"I realise how easily meanings are 'lost in translation' when discussing antennas and transmission lines; I have been in friendly discussion with a fellow club member who insists SWR is a function of line length. This is indeed the case in practice due to inevitable imbalance in most antenna systems which results in the feeder itself acting as a third radiating element to a greater or lesser extent. Thus by altering feeder length the dimensions of one (the minor) radiating element is being changed which therefore will vary the antenna feedpoint impedance and in turn the SWR. Trying to find the words to describe this in straightforward terms can be a challenge. My response to those who argue their antenna is 'good' because the SWR is nice and flat over a wide bandwidth is to point out that the system with the best 1:1 SWR bandwidth is a dummy load."

EARLY TRANSISTORS. The December 2007 review of the early development of transistors continues to stir readers' memories. John Teague, G3TGJ recalls that about 1949 while an engineering trainee at EMI, Hayes a friend, Gil Trafford, in charge of the Valve Applications Laboratory, rang him: "Come over here, I've got something to show you". G3TGJ writes: "When I arrived he revealed a

padded package he had received that morning which contained some small white-painted devices with three wires [one from each end, the third from a central rectangular block, 5 x 4 x 4mm, overall length about 15-20mm, cylindrical end tubes 2.5mm diameter]. These were the first transistors the company had obtained from RCA with which EMI was then closely connected. Their configuration was entirely different from any transistor I have seen since."

Derek Slater, G3FOZ recalls an article in 'Short Wave Magazine' in the mid-1950s by John Osborne, G3HMO, describing how to make a point-contact transistor from a germanium diode. G3FOZ writes: "I remember destroying several diodes, with inconclusive results. For me, as a chemist, the best bit was putting a point on a tungsten wire by dipping it into molten sodium nitrite. I also remember that a part of the process was to discharge a capacitor through one of the junctions. Quite why, I can't remember.'

SMALL LOOPS. I must also report receiving two letters from Canada in connection with G3LHZ's small loop claims. Alan Goodacre, VE3HX writes in support of Professor Underhill. He puts forward a hypothesis explaining why he believes it is possible to achieve high radiation efficiency. He backs this up with thermal measurements made on 14MHz with a 1m diameter copper tubing loop, comparing them with those on a 10m mobile whip. Ingenious, but one suspects that his methodology may be even more error-prone than that of G3LHZ. Dr John Belrose, VE2CV, on the other hand, strongly reiterates his considered opinion, based on theory and experiment, that the radiation efficiency claims made by G3LHZ cannot be substantiated and fly in the face of long-established theory supported by virtually all professional antenna engineers and physicists.

FINALLY. May I express my gratitude to those UK and overseas readers who in letters or during QSOs have conveyed their appreciation of TT and the pleasure it has given me. *Thanks!*

15

"Bulletin" reflections

by PAT HAWKER, G3VA

Volume 1 Number 1 of the *T & R Bulletin* was published in July 1925. Now 50 years and 600 issues later we trace just a little of its long continuous story, covering a period which has seen the hobby grow from around 1,000 experimental transmitters in 1925 to the over-20,000 British amateurs of today. For some it is hoped this account may bring back memories; for others it may serve as a reminder that even in these days of solid-state and satellite repeaters many of the ideals, the requirements, the problems and the practices of the amateur transmitter remain remarkably the same. Inevitably this is a personal story that omits much, but it is written as a grateful and frank tribute to a journal that has seemed part of the author's life for just on 40 years.

BIRTH OF THE "BULL"

By 1925 the first heady excitement of short-wave dx was over, transmitters (mostly single-stage power oscillators) and Schnell and Reinartz "straight" receivers had been made to work down to 23m; spark had been superseded; broadcasting was pushing British amateurs off 440m; aircraft communications had taken over 1,000m. Daylight dx had come with the opening of the short waves; the Radio Transmitters Society had fused with the Transmitter & Relay Section (the active transmitting group within the still prestigious, but not always effective, Radio Society of Great Britain, then concerned with many aspects of "wireless"). Radio amateurs were divided into two rival groups—the "giants" of the recent dawn of international dx and those who had helped pioneer broadcasting in Britain by phone and music transmissions on medium waves.

Until 1925 *Wireless World* had been the official journal of the Society, providing generous coverage of meetings and activities; occasionally with papers reprinted as *Proceedings of the RSGB*. There had been no regular Society publications. As interest increased in broadcasting reception, the commercial publications gradually devoted less space to transmitting topics. Several of the early British amateurs looked wistfully across the Atlantic and felt the lack of an exclusively amateur radio periodical, but (as in many other matters) it was left to Gerry Marcuse, 2NM, to make the first move.

In 1924 he wrote a letter to the editor of *Wireless World*: "Considerable dissatisfaction seems to exist in the minds of various members, the causes of which appear to vary, but the feeling is, I believe, that we require a periodical of our own, similar to *QST*."

Even then, no immediate action resulted and the idea might have lapsed had there not been a change of ownership of *Wireless World* in February 1925 (a change that arose out of the journal's support for international working by British amateurs when, in 1924, the Government attempted to close this down—but that is another story).

The new owners were reluctant to continue the official association with the RSGB and it was arranged that instead *Experimental Wireless* would become the Society's journal. This was a magazine edited by Paul Tyers, 5GX, which had supported amateur radio since the first issue in 1923, particu-

larly the "more serious experimenters". But it had never minded chastising in print those "who never perform any experimental work, who buy their sets ready made, who usually know no morse whatever, and who are best known for the great number of gramophone records which they send." Although many amateurs wrote for *Experimental Wireless*, it is doubtful whether it enjoyed the same support as *Wireless World*, so the proposed change brought to the fore the dilemma facing the T & R Section. Out-of-London members were already complaining that they received little in return for their subscriptions.

So by 1925 Gerry Marcuse and H. Bevan Swift, 2TI, then chairman of the T & R Section, felt something must be done urgently if the British amateur movement was not to split up again, as it had done several times in the early 'twenties. Bevan Swift has described what happened: "With this thought in mind, we (2NM and himself) resorted to a Lyons teashop and over a cup of coffee discussed what could be done. If we could only issue a bulletin, say once a month, detailing the activities of the section, it would give the provincial membership some satisfaction. The original idea was a simple four-page leaflet without advertisements.

"A rough draft was prepared and taken to a committee meeting of the section who immediately approved the idea and suggested that instead of a leaflet an actual magazine should be issued. As the committee was unable to embark on this expenditure itself, the matter had to go before the main RSGB Council for sanction. Here the proposal received only lukewarm support, because of the alternative arrangements with *Wireless World* and *Experimental Wireless*. The T & R Section committee decided to go ahead with the *Bulletin* and to shoulder the expense as best they could. To lighten the burden it was resolved to take a few advertisements; Arthur Hambling, 2MK, was able to approach the radio trade for advertising. J. A. J. Cooper, ex-5TR, was the first editor and with Bevan Swift, Gerry Marcuse and Ralph Royle, 2WJ, made up the first editorial committee.

"Vol 1 No 1 was hailed with general approval . . . some wanted a *Bulletin* every week and twice the size."

The first 12-page issue, of which nearly five pages were filled with advertisements, included a description of a single-stage 23m transmitter by Ralph Royle, 2WJ. The components were mounted on a wooden baseboard; there was a

The equipment used by Gerald Marcuse, G2NM, at Caterham, Surrey, during January 1924. From this station he made the first two-way contact on short waves with the west coast of the USA and maintained contact for some weeks with the Rice-Hamilton Expedition to South America

home-made blocking capacitor made of zinc sheets separated by photographic plates; the large single valve was mounted in a wooden supporting frame; the power supply used "chemical rectification" from the mains. Another article described a new "tetrodyne" superhet mixer using a four-electrode valve; there were many humorous asides which depended on the close-knit nature of the early amateurs. The *T & R Bulletin* was off to a promising start.

One result was a large increase in membership of the T & R Section which soon outgrew in size the main body of the RSGB; since members of the section were not automatically members of the main Society the anomaly led to a drastic re-organization of the whole Society: the T & R Section as such ceased to exist and in fact became the controlling factor in the Society, but the name was retained for the journal until 1942.

As many have found, it is easier to start a magazine than to keep one going, but fortunately there was plenty of enthusiasm and an obvious willingness to lend a hand. G. W. Thomas, 5YK, took over the editorship when J. A. J. Cooper had to step down. Arthur Milne, 2MI, drew many hundreds of the early illustrations. Horace Freeman of Parr's Advertising took on the problem of obtaining advertisements. Members of the editorial committee seem to have been ready to take their jackets off and tackle every aspect of production. One of these was Jimmy Mathews, 6LL, the pioneer of transatlantic working on 28MHz in 1928, who has set up a remarkable record in having, as a member of successive editorial and technical committees, been actively concerned with the production of the *T & R Bulletin*, the *RSGB Bulletin* and *Radio Communication* until finally this year he took on the only slightly less active role of a "corresponding member".

Gradually in the early 'thirties the size of the *Bulletin* increased, more rapidly after John Clarricoats, G6CL, became the Society's full-time secretary in 1932, later secretary-editor and finally editor. If this is to be a factual account, then one must admit that "Clarry" could make

enemies as well as friends. In his belief in the freemasonry of amateur radio, sometimes it seemed he was influenced as much by freemasonry as by amateur radio. But that having been said, it must equally be stressed that it was his remarkable energy and dedicated vision that in those early days converted what was still in many respects an amateur newsletter into a substantial publication of high repute. Perhaps only those with experience of publishing can appreciate just how much effort he must have put into those pre-war *Bulletins*, with the full-time help of Miss May Gadsden and the voluntary efforts of an enthusiastic group of members.

Later, in the stringent days of the second world war, he had the frustration of seeing the size fall back almost to its earliest levels with thin 16-page issues, but nevertheless an issue did appear each month, and the journal was for a time published from his home in Southgate.

GERRY MARCUSE AND EMPIRE BROADCASTING

We have noted how credit for the launching of the "*Bull*" belongs to Gerry Marcuse, Bevan Swift and the first small editorial committee. Marcuse, having conceived the idea, seems to have been willing to leave the implementation largely to others—hardly surprising in view of the enormous enthusiasm with which he pursued almost every branch of amateur activity and organization during the 'twenties.

One story, even though it is not directly connected with the *Bulletin*, is worth telling, if only to illustrate the spirit and resourcefulness of the early pioneers: the launching of Empire broadcasting by Marcuse in 1927.

This was on 32m which was at that time one of the favourite dx bands. For this was a period when amateurs picked for themselves 23, 33, 37, 43, 80, 110 and 160m for most of their operation. This was part of their inheritance based on the

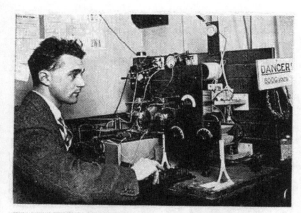

The transmitting equipment of J. H. D. Ridley, G5NN, in South Norwood, London, about 1925 when he was one of the first amateurs to hear Mexican signals on 91m. But the 6,000V and large transmitting valves show that not all stations were low power. (Photo by courtesy of *Wireless World*)

original American licences of 1912 that gave amateurs unlimited access to "200 metres and down".

Marcuse was among the most successful of the small number of British amateurs who were in at the dawn of international dx. 2NM found that his telephony operation brought him correspondence from listeners who were overjoyed to hear a voice from the "mother country". This was in 1925-26 when there were no British broadcasting stations on short waves.

And so he began, with Post Office permission, a daily series of broadcasts directed to listeners in the British Empire. And these really were *programmes*. With his friends he organized concerts and song recitals, inviting many well-known musicians and singers to his home "Coombe Dingle" in Queens Park, Caterham, and to that of a friend, Percy Valentine, in whose home a studio and control room were set up; they even had a special Post Office distribution link between the two houses in the form of two telephone lines. Valentine's uncle was a conductor and occasionally there was a full orchestra packed into the studio. Once they organized an "All-Australian" concert with internationally-known talent. There were regular outside broadcasts and the song of thrushes and blackbirds was relayed from the garden. Marcuse was not supposed to re-broadcast the medium-wave BBC programmes but he often did this, and the high-spot was his regular re-transmission of Big Ben, the first time these familiar sounds had ever been heard thousands of miles away via radio.

The Marconi Company let him have a Reisz microphone; Captain Mullard had his generator rewound when this blew up, and supplied enormous, football-sized power valves free of charge.

Marcuse had a fabulous site 700ft above sea level with a 100ft mast and Zepp aerial; he ran about 1·5kW on 32·5m and was regularly heard all over the world.

One of his first fan letters came from a lady in the West Indies. It was his lifelong claim that she wrote: "I am enchanted with your voice which I hear every Sunday morning. I have three lovely daughters and a flourishing business. If you would like to come over you can have the pick of the daughters and the business." Despite such alluring payola he remained in England.

His special Post Office licence permitted him to broadcast so many hours each week. Soon, many overseas stations (including those in Sydney, Ceylon, Singapore, India, New Zealand) were re-broadcasting his programmes on medium waves to local audiences.

These amateur broadcasts continued for about two years until in 1929, soon after the BBC had begun some experiments on G5SW at Chelmsford (it was not until 1932 that they launched the Empire Service), the Post Office told him to close down; his frequency was wanted. It had cost him some thousands of pounds out of his own pocket, but he had proved what he set out to show—that there was a demand for short-wave broadcasting from Britain. Marcuse remained an active amateur until his death at the age of 74 in 1961—a truly remarkable man who deserves to be remembered for many things, not least the foresight that British amateurs needed their own journal.

THE 'THIRTIES

It was in the years that ended with the cessation of amateur activities on 1 September 1939 that the "*Bull*" probably achieved its optimum balance and range, largely free of the economic and industrial problems that have beset post-war publishing. There were constructional articles; a constant flood of new information on aerials (some, including the "VS1AA", described first in its columns); vhf and micro-waves; design trends; workshop notes by "Shack" (G6SN); book and "contemporary literature" reviews; Contact Bureau notes (later "Research and Experimental Section"); articles for newcomers by Austin Forsyth, G6FO, and later Jerry Walker, G5JU; social happenings and the annual convention (an annual highspot from 1926 to 1938); district notes; new members and new addresses; a monthly editorial by "Clarry"; the often pungent humour of "Uncle Tom" (L. H. Thomas, G6QB). There was a neat combination of fact and fantasy in such odes as:

> "Little Tommy put a shunt
> On his meter at the front,
> If he'd hidden it away,
> He'd still have had his call today."

This was in the days when although most British amateurs were officially limited to 10 or 25W input, some had permits up to 500W or so, and fooling the PO inspector became something of a legend. Was it really true that G2 – – had a low-power transmitter in his shack with a feeder passing (apparently) straight up through the attic wherein sat an enormous amplifier? Or that the trams in Cairo used to travel in fits and starts when SU1 – – keyed his rig?

It was, however, not all fun, frivolities and "baby broad-casters" (to use a then-common term of abuse). Dud Charman, G6CJ, was showing us how to get dx to order; the careful observations on 28MHz of Dennis Heightman, G6DH, have probably never been excelled (his identification of the nature of the "hiss phenomenon" remains one of the important amateur contributions to radio astronomy); H. A. M. Clark, G6OT, was giving advice on how to avoid tvi soon after the opening of "Ally Pally" in 1936.

Print and postage were cheap and quick. The lifting of

restrictions on American imports brought in a flood of Hallicrafters, National, Tobe-Deutschmann and RME receivers and the ubiquitous 6L6, T20 and T55 valves. Components such as variable capacitors, slow-motion dials, plug-in coils were readily available from Eddystone, Raymart, JB, Hamrad, Formo and others.

A note appeared under the heading "The Month on the Air" in September 1936: "The editor is considering introducing a new monthly feature which will contain interesting news items mainly concerning dx operation. Such a feature can only be prepared by a member who is regularly on the air . . ." To this invitation John Hunter, G2ZQ, (one of those who became a "Silent Key" in the war) responded, his first column suggesting that "the stumbling block for WAZ seems to be the zone including Tibet and the Lansu province of China and we would like to hear of any *authenticated*

During the 'thirties the equipment became a lot less crude than when the "Bulletin" began, but it was still very large as this photograph shows. In the foreground, Dud Charman, G6CJ, enters up the log

contacts . . ." The column was later taken over by H. A. M. Whyte, G6WY (VE3BWY) and then Arthur Milne, G2MI, who spanned the war years.

Amateurs were on 56MHz before the end of the 'twenties (16-mile contacts were being reported in 1925) but much of the early work used simple self-excited oscillators; a big step forward was a crystal-controlled transmitter described in the "*Bull*" in November 1935. An excellent account of uhf communication, including an experimental transmitter for 50cm, was written by Eric Megaw, of GX6MU fame, in July *1931*! Regular columns on 28 and 56MHz by Nell Corry, G2YL, Constance Hall, G8LY, and others were a feature of the middle and late 'thirties. All this with a membership gradually rising to almost 4,000 by 1939. The high level of *activity* of the period is shown by the 1938 "band occupancy" checks which revealed (over a few weekends) 1,698 different British calls out of a possible 2,500. In 1938 well over a thousand were regularly active on 7MHz alone. During the period 1933 to 1938 the number of licences doubled from 1,300 to 2,600. Of 2,705 licences in October 1938, 1,812 were for 10W, 503 for 25W. There were also 2,198 "artificial aerial" licences.

Many articles appeared in the "*Bull*" on the early history of the Society—a special 21st birthday number in June 1934 ran to some 80 pages including messages of congratulation from

The Prince of Wales, Sir Oliver Lodge, Guglielmo Marconi, Dr W. H. Eccles, Sir Ian Fraser (Lord Fraser), Sir John Reith (Lord Reith) and many others. Both E. J. Simmonds, G2OD, and Gerry Marcuse, G2NM, contributed their own accounts of the pioneering of short waves. Or again, one remembers in other issues the very detailed account of "the dawn of international dx" by W. E. F. Corsham, G2UV, who took part in the original transatlantic tests, and accounts of early components by Stan Lewer, G6LJ.

ALL OUR YESTERDAYS

Throughout the 'thirties, licence conditions in the UK remained virtually unchanged (except for the reduction of the unpopular "guard bands" at the edges of each band from 25kHz to 5kHz). To obtain a "radiating" permit (full licence, minimum age 16 years) it was necessary to pass a 12wpm morse test which could usually be taken in any town (plenty of operators then to be found in the Post Offices) and also to show intention "to conduct experiments of scientific value or public utility", licences being issued "only if the nature of the proposed experiments and other circumstances warrant that course". Many and strange were the applications composed to fulfil that requirement. Perhaps it was fortunate for many of us that once the experimental licence (there were no *amateur* licences at that time in the UK) was issued, the authorities seemed to lose interest in those "experiments": though no British station was allowed to send "CQ" (we used "TEST" instead). "Artificial aerial" licences required no examinations but a full quota of birth certificates, references and the like.

Sometimes it seemed the Post Office engaged the applicant in a form of chess. The would-be amateur would submit a list of proposed experiments to improve transmitter design . . . the GPO would counter by declaring such experiments could be equally well carried out with an artificial aerial permit (which gave no right to radiate signals, merely to install or build transmitters). The frustrated applicant would consult one of the fortunate who had already obtained a licence, and together they would concoct a new thesis, this time bringing in some mention of aerials or propagation; with luck, the authority would relent and ask for "crystal certificates". If you sent only a certificate for 7MHz (doubling to 14MHz) you would get a 10W licence only for 7 and 14MHz; if you were astute enough to send also a 1·7MHz certificate, that band would be included (all other bands had to be applied for later); if you sent only a 1·7MHz certificate you got only 1·7MHz. To add 28 and 56MHz was fairly easy but 3·5MHz was difficult and the applicant had to have held a licence for at least one year. After six months, you could apply through the RSGB for a 25W permit. Higher-power permits meant that you had to dream up some more "experiments" that would justify them. The few old-timers with 500W permits were the subject of much envy. There is a classic story, again concerning 2NM, of how when his signals were picked up in Japan the Post Office (who wanted to establish a commercial circuit to that country) wrote asking him for details of the power and wavelength he was using, saying "if the limitations of your licence have been exceedeed steps will be taken, if possible, to amend the licence to regularize such tests". But that was in the 'twenties, not the more formal 'thirties!

During the 'thirties many of the more important technical developments were first applied to amateur work in the USA; this is hardly surprising since by the mid-'thirties there were already more than 50,000 *amateur* transmitters in the USA compared with under 2,000 *experimental* permits in the UK. James Lamb, W1CEI, introduced the concept of single-signal reception in 1932, putting a crystal filter in the i.f. amplifier; although his filter used a circuit developed in the UK by Dr Robinson for "stenode" reception. But the *Bulletin*, in 1939, was able to present an excellent and far-seeing series of articles by E. L. Gardiner, G6GR, (who had collaborated with Dr Robinson) on *band-pass* crystal filters. Valves were the speciality of D. N. Corfield, G5CD; book reviews were usually by Professor T. P. Allen, GI6YW. Associated publications began with *What Is Amateur Radio?* in 1932; followed by the first edition of *A Guide to Amateur Radio* in 1933, with annual editions until the appearance of the first edition of *The Amateur Radio Handbook* in 1938—240 pages and priced at 2s 6d (12½p). It was felt this was rather expensive for some prospective amateurs, so *The Helping Hand to Amateur Radio*, based on articles by Austin Forsyth, G6FO, was offered for the princely sum of 3d!

The capabilities of those pre-war amateurs should not be underestimated. After the introduction of Class B modulators in the early 'thirties, dx telephony became more consistent and led to such exploits as an all-continent round-table contact on 4 January 1938. Consider too the nine-element 28MHz rotary beam at GM6RG: six directors, driven element and two reflectors 48ft up, so sturdily constructed that the elements could be adjusted *in situ* from a gangway under the main boom: Brian Groom had this monster in use in 1938 at Galashiels.

Most of the present-day amateur contests were conceived in the 'thirties: NFD in 1933; BERU (initially a British Empire Radio Week) in 1931; top band, vhf and QRP contests were all popular, and fully reported in the "*Bull*".

Although the 0-V-1 "straight" receiver remained popular throughout the 'thirties for cw (I made my first VK contact with one in 1939) the period witnessed the great changeover to the superhet communications receiver. The HRO Senior was first marketed in 1936; National, Hammarlund, RME and the stream of low-priced Hallicrafters receivers were advertised in the "*Bull*" from about 1937. You could buy a Sky Buddy for £9 9s or a Sky Champion (with rf stage) for £15 15s, or a Super Skyrider for £32. These sets were imported by Webbs Radio, Eves Radio, Raymart, Premier and ACS. Tobe Deutschmann offered a good receiver kit, with excellent bandspread and triple-tuned i.f. transformers. Shortly before the war several British firms took up the challenge. Eddystone put their popular "All-world Two" to one side to bring out the ECR and later the 358 superhet receivers; there was a model by Evrizone (probably the first British receiver of this type, offered at £20) and the Peto Scott Trophy series.

In the early 'thirties, transmitting valves were a major problem and very costly for British amateurs. Large audio power triodes such as the PX25 were used, but they had high internal capacitances and it was not uncommon to find the old LS5 bright emitters still in use. Around 1936 the American "tubes" began to arrive: the "210" and 6L6 at about 8s 6d, the T20 at 17s 6d, the T40 at £1 4s and the 35T at £2 10s. Wonders indeed. The 807 (developed from the 6L6 about 1936) was seldom mentioned before the war. British companies brought out the ESW20 (direct equivalent to the T20 triode), and the RFP15 by the old "362" company became popular for suppressor grid modulation in 25W stations.

With coaxial cable virtually unknown, the Zepp and centre-fed dipole with open-wire lines or end-fed aerials were the most popular; the 84ft "W3EDP" enjoyed a vogue. The aerial wizards such as Dud Charman, G6CJ, talked of rhombics and W8JK driven arrays, delta matching and the importance of great circle maps. QSL cards could be bought at 1,000 for 10s from the "small ads".

WARTIME "BULLETINS"

So the 'thirties drew towards their climax. As one international crisis followed another the "*Bull*" increasingly included items about ARP, the newly-launched Civilian Wireless Reserve and its longer-established Royal Navy counterpart. Amateurs began turning up in unexpected places in unexpected roles. One read of the award of an MBE to "Tich" Emary, G5GH, for his work as a Foreign Office man in the Spanish Civil War—a foretaste perhaps of the many hundreds of amateurs who were sucked into the strange world of ULTRA and Y, double transposition and double-cross (XX), RSS and its VIs, the "Golf Club and Chess Society" at Bletchley, SCU with its farmyard and sense of discrimination, Special Forces, ISLD and Force 136, SOE and SIS. An odyssey that led one British amateur into the heart of the Netherlands' Bureau Inlichingen's secret communications network in a Dutch museum and waterworks. Although the Society was often the vital link that directly or indirectly put members in touch with such organizations, little of these activities appeared in the security-conscious columns of the wartime "*Bull*".

The blow had come suddenly but not unexpectedly. A notice in the *London Gazette* of Thursday 31 August 1939 proclaimed:

"I, Major The Right Honourable George Clement Tyron His Majesty's Postmaster General hereby give notice that . . . all licences for the establishment of wireless telegraph sending and receiving stations for experimental purposes are hereby withdrawn."

Spot the "73" and the HRO receiver in this wartime greetings card

"Belgian meteo service for the RAF"—one of the courageous "underground" organizations that sent regular weather reports into England by secret hf radio links during the second world war, so helping to reduce RAF losses due to unexpected bad weather

"*Bull*" cartoons showed the desolate amateurs watching as their equipment was trundled off into storage by Post Office officials.

"The Month on the Air" became "The Month off the Air". Members of CWR became the "Early Birds" (Wireless Intelligence Screen) etc. Amateurs were soon to be found in every nook and cranny of the world's first electronic war. Industry adapted quickly; it is said that Ernie Dedman, G2NH, of QCC, more used to supplying crystals to amateurs, found himself one day with an order for millions!

At first, paper remained plentiful and issues substantial. "Khaki and Blue", "Ham Hospitality", active-service lists and similar columns reflected the change from civvies to uniform. For all too many, a "Silent Key" notice or years in the "Kriegies" (POW camps) where more than one amateur participated in the building and operation of secret radios, and for whom C. H. L. Edwards, G8TL, organized an efficient service of gift parcels (not all of which, I suspect, were as innocent as they seemed).

For many, in the tedium that accompanies war, the monthly arrival of the "*Bull*", no matter how thin it became, was a welcome link with the hobby they had left behind—or, increasingly as new members came flooding in after *The Amateur Radio Handbook* became virtually an official Services textbook, a hobby to which they looked forward. Even from thin issues one could learn of the growing importance of frequency modulation and microwaves.

Towards the end of the war, with hostilities ceasing in Europe, one began to hear G7 callsigns on the amateur bands, operated by a small group of well-known British amateurs who spent their time working a strange assortment of stations that often concealed the identities of other amateurs overanxious to re-engage in a hobby that had been almost silent for so long (never in fact completely silent, since the Germans as part of *their* signals intelligence maintained some D3 and D4 stations on the bands, including a series of beacon stations).

The authorities, however, played fair. Licences with the unrestricted right to call CQ began to be re-issued at the beginning of 1946. My first post-war entry in the log reveals a 28MHz contact with SV1EC in Athens (the operator was Major-General Eric Cole, better known as SU1EC and G2EC, which was the only British two-letter callsign containing an E). It was not long before the first of the new G3-plus-three calls was heard; pre-war holders of figure-two-three-letter "artificial aerial" licences, on passing the morse test, were awarded their G prefix. Now for the first time there was a Radio Amateurs' Examination, though at first many could claim exemption on the grounds of Service qualifications. The terms of the new *amateur* licence had been agreed with the Society during the war.

The shortage of paper and the small 6pt typeface used for many of the pages made the early post-war "*Bulls*" a mere shadow of the fat issues of 1939: yet they included such important reports as those from Atlantic City, where RSGB representatives Stan Lewer, G6LJ, and John Clarricoats, G6CL, succeeded, among other things, in adding the vital footnote to the Radio Regulations that left us with 200kHz of "top band".

Many, better left untold, stratagems were used to obtain extra paper, and for a period an occasional *Proceedings of the RSGB* helped relieve the acute technical starvation of the period. It was the 'fifties before there seemed sufficient space really to reflect the many changes that were fast galloping into amateur radio. But one gem of a series in the late 'forties was "In Your Workshop" by "Donex" (Ken Alford, G2DX, who first held a licence pre-1914 and who must be one of the very few people whose amateur activity already spans more than 60 years).

My own professional connection with the *Bulletin* lasted from 1947 to 1951, though some readers seem to think I am still on the staff! David McIlwain and then John Rouse followed me as assistant editors.

THE SECOND 25 YEARS

To this writer at least, it seems incredible that there have been as many issues since 1950 as there were before; the years appear to have speeded up and one can never hope to do justice to the hundreds of members who have contributed to its columns. Though it is invidious to name names, one remembers Louis Varney's (G5RV) efforts to show us how to build transmitters that radiated fewer harmonics. For undoubtedly the reduction of tvi (and nowadays afi) has remained throughout this period the single most-worrying aspect of amateur radio, and one that affects directly almost all transmitting amateurs: some excellent articles have helped us all.

Then again there has been the consistent encouragement given to vhf and microwave operation—among those who have sat in the seat of vhf reporter was, for a ten-year stint, Fred Lambeth, G2AIW, and, not far short of that number, Jack Hum, G5UM, (who in the early 'thirties had contributed regular notes on 2MHz operation!). It is already some five years since Dain Evans, G3RPE, began the first monthly microwave column in any amateur journal.

There have been regular columns on ssb, on mobile operation, on amateur television, and currently on Raynet and swl activities. "Month on the air" has proved the longest-running of all regular features. One remembers with gratitude the re-awakening by Dick Thornley, G2DAF, of interest in the home-building of high-performance receivers and ssb transmitters, an early (1951) introduction to switched wide-band exciters by Reg Hammans, G2IG, and some notable contributions by Peter Martin, G3PDM. In a world increasingly dominated by speech there was a memorable account of the morse code and morse keys by J. Piggott, G2PT, in 1956. Then there have been innumerable reports from readers and from overseas technical journals in "Technical topics", which I began with more than a few doubts in 1958, but which has somehow survived 17 years under the same management—from 6ft racks to "table-top" transmitters, transistors, FETS and integrated circuits.

John Rouse, G2AHL, succeeded "Clarry" as editor in 1963 and worked diligently and skilfully to improve the journal until his untimely death in 1967. Trevor Preece, G3TRP, and John Adey valiantly kept the issues coming until the appointment, late in 1969, of our present editor: A. W. Hutchinson. Under his guidance *Radio Communication* has reached a certified ABC figure of 17,816 copies per issue; has topped 100 pages in a single issue; and has maintained consistently high standards of accuracy and detail. Advertising, with the aid of C. C. Lindsay, has for the first time since the earliest days been taken back in-house with outstanding success. Over many years Derek Cole has provided thousands of complex drawings from rough sketches with an expertise and promptness that few other journals could hope to equal.

Behind the scenes for many years has been Roy Stevens, G2BVN, chairman of the Technical & Publications Committee, who has shown how firmly he believes that committees are for *doing*, not just for discussing. It is not always realized that all technical contributions go to referees for evaluation; usually members of the T & P Committee, but often outside specialists. The role of these members in maintaining the high reputation for technical accuracy and judgement should not go unacknowledged. George Jessop, G6JP, has long been associated with the journal and many other RSGB publications.

The *RSGB Bulletin* became *Radio Communication* in January 1968—a change that at first seemed not to appeal to the majority of members. Nowadays, however, and particularly since one began to hear the diminutive "*Rad Comm*", the present title is so well accepted that, at a recent AGM, members were overwhelmingly in favour of its retention.

A society journal is not, of course, just a technical magazine. It must reflect all interests and activities of members. The more amateur radio polarizes or coalesces into factions of specific interest with mode or band rivalries, the more difficult becomes the task of the editor. The man who wants to build a top band rig is not going to seek help from articles on microwave plumbing; a channelized nbfm 144MHz mobile rig has something, but not all that much, in common with an hf ssb transceiver. Again some activities now seem to engender a desire for exclusiveness and the channelling of information into specialized newsletters.

Constructors want constructional articles; buyers want equipment reviews; some want controversy—some wish to avoid it; a local group or society expects to see its activities reported, even though these may be of local interest only; some would like pages of "Your opinion"; almost all readers want pages of ads and small ads. Some want more space given to this or that; others resent any space being given to activities in which they personally are not interested.

The greater diversification of amateur interests, and the growing gaps between them are, I believe, the main reasons why one can look nostalgically at those *Bulletins* of the middle and late 'thirties and feel that they had perhaps more central unity and balance than has generally been achieved in the second 25 years. Compared with 40 years ago, there is these days much less humour, a noticeable absence of sermonizing or policy-stating editorials (and sermons are good for all of us sometimes), less looking towards the future but also less looking back at the lessons of the past. It seems sad that a Society five or six times as large has not been able to keep an annual convention going, as it did so successfully in the 'thirties. Even an annual exhibition is no longer directly linked to the RSGB. The many specialized factions and newsletters are symptoms, not the cause, of this difficult problem. There is a critical size to an amateur radio society above which it becomes increasingly difficult to maintain personal contacts and enthusiasms.

Of course, today it is more streamlined, efficient, cost-effective and professional and few of us would have it otherwise, whether or not we question the idea that "bigger necessarily means better". But I must confess that personally when, for any reason, my belief in the future of the hobby occasionally falters, I find it soonest restored by picking up some old *Bulletins* (thanks to Reg Cole, G6RC, my shelves go back to 1928, some eight years before a 14-year-old schoolboy became hooked on his own copies). That perhaps is the nature of the ageing process. I hope and imagine that in the year 2000 the new members of today will be looking back with similar warmth to Volume 51.

Happy birthday, O journal, and I hope that in your veins of printer's ink you feel 50 years *young*! □

HF BEACON STATIONS

Callsign	Frequency (MHz)	Location	Reports to
DL0IGI	28·195	Mt Predigtstuhl near Salzburg	DJ5DT, Kollowitz-weg 1, D 6100 Darmstadt, FR of Germany
GB3SX	28·185	Crowborough, Sussex	G3DME
PY1CK	28·165	Rio de Janeiro	PO Box 1044, Rio de Janeiro, Brazil
VE3TEN	28·175	Ottawa	VE3QB, 782 Dunloe Avenue, Ottawa, Ontario, Canada
VP9BA	28·165	Southampton, Bermuda	VP9BY, PO Box 73, Devonshire, Bermuda
ZL2MHF	28·170	Mt Climie, Wellington	PO Box 40212, Upper Hutt, New Zealand
5B4CY	28·180	Limassol	5B4AP, Box 1267, Limassol, Cyprus
3B8MS	28·190	Signal Mount, Mauritius	MARS, PO Box 13, Curepipe, Mauritius

Reports for any of the above may be sent to RSGB HQ (Attn IBP). At present only DL0IGI switches to 28·200 at 00-05 and 30-35min past each hour.

Questions and Answers

RadCom readers were asked to put questions to Pat about 'TT' and his life in publishing and radio. Below you will find them, plus his answers.

How has the RSGB as an organisation contributed to the science of radio?

Basically, by the publication of its journal. In the days of Dr Smith Rose, by setting-up satellite committees. The society made a significant contribution to the International Geophysical Year (1957-1958), which was set up very much in conjunction with RSGB.

Have you ever written under another name?

Yes. 'The Future of Amateur Radio' (by 'Navigator'), which was published in *Wireless World* in February 1941 was Pat's first *nom de plume*. It was a feature that produced a heated, angry response from John Clarricoats, G6CL, the General Secretary of RSGB, in the following edition, plus a *Wireless World* editorial, plus several letters to the editor. It was many years before Pat would admit to being 'Navigator', a pen name he had coined for a direction finding article he wrote for *Armchair Science*.

So why did such a well-known and highly respected author choose to write under another name? In some cases it is because Pat didn't want readers to know who he was. He wrote a couple of pieces under the name Patrick Halliday for *The Guard-*

ian, and quite often used that name in *Electronics Weekly* because in lots of magazines they don't like having two feature articles by the same person in one edition. Halliday was Pat's mother's maiden name.

Pat also wrote under the name Patrick Ross, Ross being his son's middle name.

What has been the most significant development in radio, made by amateurs?

Not many significant developments have been made entirely by amateurs, but amateurs were responsible for quite a lot of developments in the course of their professional careers in communications. These were then brought into amateur radio.

The Moxon Rectangle, which was based on the VK2ABQ wire beam is an example. The VK beam was end coupled and was an amateur development, but Pat is not sure if it ever passed into professional use.

An amateur development that has certainly made the crossover into professional use is the H-mode mixer, developed by Colin Horrabin, G3SBI. It is now appearing in some Japanese and also in some military equipment. Having said that, G3SBI is a professional engineer, but he developed the H-mode mixer purely as an amateur enquiry. The first commercial receiver to incorporate it was the AOR AR-7030.

Pat feels that some 'TT' readers may have become bored of the H-mode mixer by now, as it is still being developed and is mentioned frequently, but he believes it to be an important contribution to HF receiver design.

Which amateurs have contributed most to the science of radio?

If you go right back, practically all the major developments, especially the American ones, came from amateurs who had become professionals. One of the things that Pat used to notice when he attended conferences is that with American speakers the synopses attached to papers would frequently mention the fact that they were - or had been - amateurs, whereas the British engineers seldom did because there was an overwhelming feeling on this side of the Atlantic that 'amateur' meant 'non-professional'.

At one time, many years ago, Martin Sweeting, G3YJO – now

Professor Sir Martin Sweeting OBE FRS, the Director of the Surrey Space Centre as well as CEO of Surrey Satellite Technology Ltd – did some development work that was written up by his professor for publication in *Wireless World*. The professor carefully avoided using the word 'amateur' in any sense of the word, which left Pat feeling extremely indignant. His response was to write a piece for *Wireless World*, pointing out that that it was largely an amateur development. The professor responded, pointing out that he had indeed avoided using the word 'amateur', because it could be misinterpreted.

Professor Sir Martin Sweeting OBE FR, G3YJO.

Irrespective of whether they acknowledged that they were radio amateurs or not, Pat says that at a conference you could usually tell by the technical vocabulary whether someone was licensed or not.

With modern technology becoming ever more complex, do you think there is still an experimental role for amateurs to play?

Yes, but mostly by professional amateurs. Certainly there is still useful research to be done in the field of aerials, but not all of what is discovered is useful or actually true.

Martein Bakker and others are doing extremely good work on receivers using the Internet to exchange information. They are still looking for the perfect receiver, but Pat feels they ought to concentrate a bit more on the ideal transmitter, because there's no point in having a receiver with extremely low phase noise if the transmitters have 'noisy' frequency synthesisers.

Dick Rollema, PA0SE, who Pat has exchanged numerous messages and ideas with.

The ARRL have largely stopped presenting experimental design work in *QST*, but it continues very much in *QEX*, also published by ARRL. Pat recalls that there is a "terrific controversy" in there

at the moment on the subject of a completely new transceiver that uses the H-mode mixer. There are claims that it is by far the best transceiver ever designed by amateurs. G3SBI and others in the group are a little dismayed at this because they claim that the CDG2000, published in *RadCom* in 2002 is still the best. The CDG2000 was designed by Colin Horrabin, G3SBI, Dave Roberts, G8KBB, and George Fare, G3OGQ, each or whom gave the first letter of their names to the model number, hence 'CDG'.

Pat has some interesting statistics on the construction of the CDG2000. About 250 people are said to have embarked on building one, about 100 in Britain and 150 overseas. However, some of the components were no longer being produced, even within a few months of the feature appearing. It all goes to show just how fast technology is marching on and some components are becoming obsolete. This is in stark contrast to the valve era, when once a particular valve went into production you could be pretty sure of getting one for the rest of your life. These days ICs have a short production life. The Plessey SL600 linear series of integrated circuits is a case in point. Very popular with home constructors, when Plessey sold out to GEC they simply stopped making them. The 600 series are still obtainable from a few sources, including the G-QRP club. This is a general trend. Within a few years of a design appearing, you can't make it any more. Moreover, if anything goes wrong you can't get a spare. This is just as true with Japanese factory-built equipment.

Has Software Designed Radio opened-up avenue for experimentation by amateurs?

At the moment you've either got a software engineer involved or an RF engineer who designs the front end. At the moment the front ends are still basically the same as front ends for conventional equipment. Pat feels that all new amateur transceivers are going to have some DSP in them, but whether they have a stand-alone DSP or it is linked to a personal computer is another matter. Pat admits to having gotten into an argument with the British agents for Flex Radio, because his knowledge – gained from well-informed sources – is that the design of their original front end was not very good. Also, the digital board was placed too close to the front-end board, resulting in noise being transferred to the front-end board, raising the noise floor. The original designs were really only good up to about 14MHz, but progress is being made. When it becomes possible to convert to digital at the first IC, it won't be within the means of amateurs to do anything, so at the moment it is a good field for experimentation. It certainly has a future.

Modern PCs, which are used to handle digital modes via the sound card, have had a big impact on amateur radio, especially the ability with some modern datamodes to be decoded even though the signals cannot be heard. Pat points out that even this is not entirely new, because the old Piccolo system, in use from 1963, was advocated on the sense that it could be received 4dB below the noise level. Synchronous systems are the key, because with such systems it is possible to anticipate what is going to be received. Having said that, Pat prefers to hear a station, rather than just see the copy on screen. Pat is even less than 100% sold on SSB, because he sees its predecessor, AM, as a fine recruiting tool for youngsters who listen on the HF bands with a broadcast receiver and might come across a station they could resolve.

For the last 35 years or so, the Japanese have dominated the Amateur radio market. Perhaps the arrival of Software Defied Radio will restore a balance. Certainly it is interesting to see that Motorola has taken over Yaesu, stating that production for the amateur radio market will continue.

How much has the miniaturisation of components contributed to the reduction in home construction these days?

As far as Pat is personally concerned it's a great deal, as he finds it difficult to handle an ordinary Integrated Circuit, let a lone a surface mount IC. More and more, unless someone has very good training or has an extremely steady hand, it is just not possible to make things with the latest generation components. On the other hand, for people who want small things, it's a boon, although Pat cannot see the need to make everything small and portable.

Pat's personal preference is for large equipment, but acknowledges that this doesn't suit everybody. The reason for this is that large equipment is easier to repair. Contrast this with the latest generation transceivers, which hardly anyone would attempt to repair at home.

There are very few components advertised in *RadCom* these days and it isn't possible to re-use many components from commercial equipment, so it is viable for amateurs to develop their own equipment any more?

There are still plenty of full size components around. You still see components advertised in *Practical Wireless*, and indeed valves, so obviously these things appeal more to the readers of *PW* and *Sprat* (G-QRP Club).

What about antennas? Is there still useful work to be done in that area?

Certainly, but then again, because of NEC, everybody simulates them before making them. But simulation is not entirely satisfactory when people live in restricted spaces, with houses and trees close by. The slight difficulty with experimental antennas is that in the old days we just used to put up a bit of wire and hope for the best, whereas nowadays it is necessary to have some knowledge about the fundamentals of radiation in order to distinguish the difference between something that's new and something that isn't.

So far as Pat is concerned, the biggest antenna controversy in recent years has been all about electrically short antennas, including loops. When GM3HAT came out with his E-H theory, published first in *Wireless World*, it was immediately pounced on by a number or people who pointed out that what he said was not the case and that separate radiation of the E and H fields just didn't make sense as they are inextricably linked.

That controversy has continued ever since and Pat now refrains from commenting on the subject. But then Professor Mike Underhill came along with his theory on short loops. Because of his qualifications it is a little difficult to challenge him, although Pat doesn't agree with his theory. To put that into perspective, Pat has great admiration for Mike's work some years ago on oscillators, adding "As a Mullard design engineer he did some very good work."

Some critics of 'TT' complain about the column constantly revisiting the past. Do you think that looking back inhibits or promotes the development of amateur radio, new ideas and new techniques?

Pat has always believed in looking back, even in the first 'TT'. The reason for this is that there are always new people coming into the hobby, and Pat's feeling is that many of those new people don't have the background. Unless you know what *was* and what wasn't developed previously, and what went wrong and what disadvantages there were with the older equipment, you're not really in a position to improve the present equipment. In any case, history is fundamental to any subject. In many ways

Pat is more interested in the historical side than the development of new things.

Some *RadCom* correspondents are professional amateurs. Do feel that being amateurs helped their professional work and vice versa?

Certainly! RF engineering has gone out of fashion at universities and so on, so it is only via amateur radio that you can interest people in the subject nowadays. Pat recalls that a few years ago a university lecturer quizzed students on how mobile telephones worked. Apparently only a couple realised that there was a radio link involved in the process. In these 'digital' days, people just don't think of it being associated with radio engineering, but the analogue side of radio engineering is still important. It may disappear one day, but even then broadcasting will still rely on high power transmitters.

In this day and age, when the vast majority of people operate commercially-made equipment, what justification is there for insisting that those taking any radio communication exam – particularly the Advanced one – have a knowledge of how their equipment works?

Pat's justification is "Do you want Citizens' Band?". The reason for CB is that there is no need to take an exam. You just buy the equipment and operate it. In many ways it was a good system, but it was ruined by the people who jumped into it and didn't regard radio as anything other than a means to communicate.

There have to be standards, and the only way to differentiate between CB and amateur radio is to understand the technology. If all exams were done away with, except for knowing the regulations, a lot of people wouldn't stay in amateur radio for longer than a few months. If you don't make some sort of commitment to a hobby, it doesn't usually last.

A lot of young Americans have found that they take out an operators' licence and never get on the air. It must also be true in Japan, where there are – or were – about two million licensed amateurs. But there aren't two million on the air – or at least it doesn't sound like it.

Do you ever – or did you ever – take part in contests?

Pat has taken part in innumerable contests, from Field Days when amateur radio was reintroduced after WWII, right up to the present day. Pat is a regular participant in RSGB's Commonwealth Contest (BERU) and ARRL contests, although these days he doesn't submit entries because he treats them as transient events. The author enjoyed a completely unscheduled contact with Pat in an RSGB Club Championship CW contest as this book was being written.

What periodicals do you regularly read?

In short, not the same ones now as years ago. Pat has always been a big user of libraries, but these have changed over the years and amateur radio is now dropping out of the conscious-

Photo: Julian Greenberg, G4ZOD.

The IET Library, in Central London.

ness of professional librarians. As Pat sees it, the only library that is good for amateur radio these days is the IET (Institution of Engineering and Technology) Library, located at Savoy Place in Central London - 'IET' being the new name for the IEE (Institution of Electrical Engineers). It is the only library Pat knows that displays a copy of the ARRL magazine *QST*. It also takes a lot of IEEE professional publications, which Pat finds very useful. Pat doesn't try to look at all of these, but he does consult the IEEE Transactions on Antennas and Propagation. The IEEE also has a separate magazine on antennas and propagation, which is useful.

At home, Pat reads *Practical Wireless*, *QEX*, the G-QRP Club magazine *Sprat*, *Radio Bygones, OTNews*, *Mercury* and the British Amateur Television Club magazine *CQ-TV*. *Electronics World* is consulted at libraries, as is *New Scientist*. *Television and Electronics Consumer* used to be used, but that disappeared and has now been revived as a web-based magazine called *Television and Consumer Electronics*. Pat admits that he is about to stop his subscription to *Time* magazine, because he doesn't have time to read it. *The Guardian* has some useful technical content, but – along with many other newspapers – has become so large that it would take all day to read.

While at the IBA, Pat used to see the European Broadcasting

Union technical journal. That also used to appear in the British Library, but these days you have to order the magazine and Pat much prefers to just browse.

RF Engineering for Experimenters is a book that Pat finds useful, indeed it covers the subject of the H-mode mixer. The latest editions of RSGB's *Radio Communication Handbook* also contain it.

Do you believe there was a golden age of amateur radio? If so, when was it?

Pat believes that every amateur has his/her own golden age of amateur radio. Every amateur starts as a great enthusiast. The first contacts made or the first equipment built and made to work constitute a golden age. As to whether things have gone downhill or not, that's another matter, but Pat is optimistic that amateur radio will survive.

What are your views on Morse?

Pat is all for it, and feels that the Morse test which has now been dropped as a requirement was extremely useful. The reason for this is that the Morse test made sure people were committed to the hobby. Pragmatically, Pat sees the dropping of the Morse test as a necessary thing. "The time had come for it to be dropped", he says. Technically, with so many modes available for amateurs to use now, it was no longer justifiable to put one mode above all the others. Pat remembers that before WWII people didn't have to use Morse, indeed a few didn't pass a test. A licensed amateur could in fact have a Morse operator for his amateur station. After WWII the situation was turned around completely, because newly licensed amateurs were required to spend a year on Morse before they were allowed to use telephony. Pat's first contacts were most definitely on AM telephony.

Pat would encourage everyone coming into amateur radio to learn Morse, but not necessarily before getting a licence.

Has amateur radio been dumbed-down in recent years?

Pat sees 'dumbing-down' as a difficult question. For economic reasons it is necessary to attract youngsters into the hobby, but these days it is very difficult because they are more likely to be attracted by Information Technology and digital systems than amateur radio. This is nothing new, because Pat tried unsuccessfully to get his son Philip to learn Morse. "Even he went into IT and has stayed there ever since!", says Pat, adding "Philip is currently Communications Network Manager for British Airways - a company that has recently handed over its HF communications to a Swedish company."

Finale

By 2004, Pat's wife Gwen was clearly showing signs of Alzheimhers. Following a bad fall she could no longer look after herself, so she returned to Dovercourt Road where Pat looked after her for the next 22 months. When she went to hospital in September 2006 it was decided that she needed attention and facilities that could not be provided at Dovercourt Road, only at a nursing home where she now lives. Pat, himself with arthritis-limited mobility, makes weekly visits, and there are fairly frequent visits by their daughter Virginia, with a career in Social Services, and by their son Philip, husband of Jane and father of Pat's four grandchildren.

Pat is once again living on his own. When in his shack he is surrounded by equipment that he knows and understands, his impressive library of archive material close at hand. Although not as mobile as he once was, his mind remains as sharp as ever. With his ageing KW2000A he remains active on the HF bands and even takes part in contests.

Pat admits to having been lastingly influenced and affected by his wartime involvement with secret intelligence, the friends he made, and the respect he still feels for the clandestine agents who operated HF transmitters in enemy-occupied territory. He still makes an annual trip to Bletchley Park, to attend the RSS and SIS Section VIII reunions organised by Bob King, G3ASE. Until his death in 2001, Pat kept in touch with, amongst others, Watson ('Bill') Peat, CBE, GM3AVA, who became a prominent figure in Scottish agriculture, various business enterprises and for five years a Governor of the BBC.

A Bit of Controversy

It is only right and proper that in concluding this book I offer my sincerest thanks to Pat, who couldn't have been more helpful in the research for it. He provided not only a lengthy personal interview, but access to numerous rare documents and unique photographs. A quiet, modest and reflective man, over the years Pat has brought pleasure to many by writing not only 'Technical Topics', but writing features for other magazines, and editing books and magazines for other organisations. Naturally enough, not everything he has written has found universal favour, but as they say, 'You can't make an omelette without cracking a few eggs'. And Pat really has been a most prolific author and disseminator of information.

I will leave the penultimate words to Pat himself. Taken from his '50 Years Have Slipped Away' feature for the 50th anniversary issue of *OTNews*, the journal of the Radio Amateurs Old Timers' Association, he said "Whether you feel that the hobby has improved or deteriorated since 1958 is a matter for each of us to decide in our own way, but it is all of our responsibilities to see that it continues to flourish as a responsible and self-training hobby."

He'll be a tough act to follow.